Managing the Future of Southeast Asia's Valuable Tropical Rainforests

Advances in Asian Human-Environmental Research

Aims and Scope

The series aims at fostering the discussion on the complex relationships between physical landscapes, natural resources, and their modification by human land use in various environments of Asia. It is widely acknowledged that human-environment-interactions become increasingly important in area studies and development research, taking into account regional differences as well as bio-physical, socioeconomic and cultural particularities.

The book series seeks to explore theoretic and conceptual reflection on dynamic human-environment systems applying advanced methodology and innovative research perspectives. The main themes of the series cover urban and rural landscapes in Asia. Examples include topics such as land and forest degradation, glaciers in Asia, mountain environments, dams in Asia, medical geography, vulnerability and mitigation strategies, natural hazards and risk management concepts, environmental change, impacts studies and consequences for local communities. The relevant themes of the series are mainly focused on geographical research perspectives of area studies, however there is scope for interdisciplinary contributions.

For further volumes:
http://www.springer.com/series/8560

Managing the Future of Southeast Asia's Valuable Tropical Rainforests

A Practitioner's Guide to Forest Genetics

Edited by

Ratnam Wickneswari

Universiti Kebangsaan Malaysia, Bangi, Malaysia

Chuck Cannon

Xishuangbanna Tropical Botanical Garden, Menglun, China

Springer

Editors
Ratnam Wickneswari
Universiti Kebangsaan Malaysia
School of Env. & Natural Resource Sci.
Faculty of Science & Technology
Bangi
Malaysia
wicki@ukm.my

Chuck Cannon
Chinese Academy of Sciences
Xishuangbanna Tropical
Botanical Garden
Menglun
China, People's Republic
chuck@xtbg.ac.cn

ISSN 1879-7180
e-ISSN 1879-7199
ISBN 978-94-007-2174-6
e-ISBN 978-94-007-2175-3
DOI 10.1007/978-94-007-2175-3
Springer Dordrecht Heidelberg London New York

Library of Congress Control Number: 2011934881

Printed on acid-free paper

Springer is part of Springer Science+Business Media (www.springer.com)

Foreword 1

On behalf of Bioversity International, I congratulate Prof. Ratnam Wickneswari and Chuck Cannon, co-editors of this book, *Managing the Future of Southeast Asia's Valuable Tropical Rainforests: A Practitioner's Guide to Forest Genetics*. I also congratulate the other contributors to this book.

The publication of the book is very timely. It gives deeper meaning to the celebration of the UN International Year of Forestry that we are observing this year. We are also in the process of coming-up with a review of the state of the world's forest genetic resources. I am sure that the book will give more meaning to our celebration and further enrich and provide greater depth to our understanding of the state of forest genetics in Southeast Asia.

I find the book to be reader-friendly. Indeed, it was written with the forestry students, forest managers, policy makers and general readers in mind. In terms of content, the book provides information not only on forest genetics of the Southeast Asia's tropical forest as the title indicates but also on a very interesting historical (including biogeography) perspective to genetic diversity of the forest trees and forest animals. I find many of the information provided to be very useful and enlightening.

The book offers more. It provides a practical genetic consideration to ensure sustainable forest management. According to the authors, this is very important in reducing the impact of logging practices to the environment.

As you may be aware, one of Bioversity International's focus is on the conservation and use of plant genetic resources, including forest genetic resources. Hence, Bioversity International has been actively engaged in activities relevant to the conservation and use of forest genetic resources in the region. We work closely with the Asia Pacific Forest Genetic Resources Network (APFORGEN) which is associated with the Asia Pacific Association of Forestry Research Institutions (APAFRI). APFORGEN was established in 2003 with the general objective of making countries in the Asia Pacific region to be more coordinated and committed in the conservation, management and use of forest genetic resources. Another important aspiration of APFORGEN is to facilitate easier access to forest genetic resources information in the region.

I am sure that our partners in AFORGEN and APAFRI, like Bioversity, will be delighted to have this book that showcases the different works done in this region.

This book is a must read for practitioners and those interested in learning more about forest genetics in the context of biogeography and forest management in Southeast Asia.

Regional Director,
Bioversity International Asia,
the Pacific and Oceania,
Malaysia

Leocadio S. Sebastian

Foreword 2

Southeast Asian forests are largely known for their unique and highly diverse biological resources, valuable timber and non-timber species and natural products of medicinal value yet to be exploited. Sustainable management of these forests has become an international agenda since the signing of the Convention on Biological Diversity in Rio de Janeiro 1992, requiring actions at the regional and national levels.

International and national funding for research on biodiversity and development of local expertise have been instrumental for the current sizeable number of scientific publications on genetics, reproductive biology and systematics of the flora and fauna found in Southeast Asian forests. These new findings need to be incorporated in forest conservation and management guidelines for sustainable timber production from natural forests whilst ensuring conservation of biodiversity.

This book provides a detailed overview of the evolutionary biology of tropical rainforest trees and how management systems interact with these natural dynamics. The authors have provided an up-to-date and insightful review of important scientific findings and conclude with practical recommendations for the modern forester in Southeast Asia.

This book will prove beneficial to forestry students, professional forest managers and policy makers, who do not have technical training in genetics. It is also intended for non-specialists who are involved in the tropical timber industry, from the local forest manager to the international timber purchasing agent.

Science Advisor to the Prime Minister of Malaysia Dr. Zakri Abdul Hamid

Foreword 2

Foreword 3

The forests of Southeast Asia have long been valued as source of food, fuel, fibre and medicine. It will continue to play an important role in rural economies and influence the socio-economic development of the respective countries and finally the quality of life. Asia and the Pacific region account for about 70% of the world production of tropical industrial roundwood. Hence, how are we going to manage these forests sustainably so that our children, grandchildren, great grandchildren and generations to come can continue to enjoy the good quality of life? Moreover, with the increasing global concerns on the environment and especially on climate change and biodiversity conservation, the importance of good forest management cannot be underestimated.

As a forester I would like to reinforce the statement by D.M. Smith in 1962, 'a good silvicultural system is not chosen but formulated as a solution to a specific set of circumstances'. Criteria and indicators for sustainable management of forest and conservation of biodiversity have been developed in most countries in Southeast Asia. A practical guide which incorporates genetic conservation in the various stages of activities undertaken during forest harvesting needs to be updated regularly as our understanding of marco- and microevolution of tropical forest trees increases.

This book which addresses these pertinent questions and provides solutions to management of tropical forests is a worthy read to forestry students, professional forest managers, policy makers and non-specialists who are involved in the tropical timber industry. I would also like to congratulate Prof. Dr. Ratnam Wickneswari for her initiative.

Academician Tan Sri Dr. Salleh Mohd Nor
Secretary General,
Academy of Sciences Malaysia
Council 2011–2012

Preface

Technical publications on genetics, reproductive biology, ecology and biogeography of Southeast Asian forests have seen an increase in the past decade largely due to development of expertise in the region together with interest and concerted involvement from the international scientific community on research in the region's biological and natural resources. Funding from national governments and international agencies to undertake these studies on biodiversity using new technologies was very forthcoming and led to many scholarly publications.

This book was conceived about 3 years ago taking into consideration our combined involvement in tropical forest genetics research for about 30 years and the good number of publications in the area. We then agreed upon a conceptual framework and identified key contributors for the book based on publication track record and experience in tropical forestry. When the book was about 70% completed, we pitched for funding to publish the book and secured a publisher who greatly recognised our scholarly work and implications to conservation and management of tropical forests in Southeast Asia.

Through this book we wish to provide an accessible statement on the current state of knowledge about tropical forest genetics and its implications for the profitable and sustainable management of forest resources in Southeast Asia. This book will prove beneficial and is aimed at forestry students, professional forest managers and policy makers, who do not have technical training in genetics. It is also intended for non-specialists who are involved in the tropical timber industry, from the local forest manager to the international timber purchasing agent.

Each chapter will cover a major topic in the evolutionary biology of tropical rainforest trees and how management systems interact with these natural dynamics. Authors will provide an up-to-date and insightful review of important scientific findings and conclude with practical recommendations for the modern forester in Southeast Asia. Several chapters will provide compelling discussions about commonly neglected aspects of tropical forestry, including the impact of historical dynamics of climate change, anthropogenic threats to genetic viability and the important role of wildlife in maintaining genetic diversity. These discussions will promote a deeper appreciation of not only the economic value of forests but also their mystery and intangible values.

The timber industry in Southeast Asia is a major contributor to the regional economy but the connection between scientific research and the application and development of policy is weak. This book will help bridge the knowledge gap between academic publications and forest practitioners. Besides, this book will be the first of a kind that reviews the Southeast Asian forest genetics and evolutionary biology. In addition, it will also serve as an excellent reference book for evaluators of Forest Stewardship Council certification. We could also say that our book is an overview of conservation and management issues in tropical forestry.

The language used is relatively free of jargon. Chapters are relatively brief focusing on a clear outline-like structure and are free of verbose explanations. They are structured like an invited lecture to a general audience of knowledgeable people. Creative and simple illustrations are included where necessary.

We would like to thank the lead authors of chapters for their commitment and timeliness. Support from Universiti Kebangsaan Malaysia and Texas Tech University is gratefully acknowledged.

We hope this book will serve its intended purpose now and in the future.

Bangi, Malaysia — Ratnam Wickneswari
Menglun, China — Chuck Cannon

Contents

Contributors

Richard T. Corlett Department of Biological Sciences, National University of Singapore, Singapore 117543, corlett@nus.edu.sg

Mohamad Ismail Shaharuddin Institute for Environment and Development (LESTARI), Universiti Kebangsaan Malaysia (UKM), 43600 UKM Bangi, Selangor Darul Ehsan, Malaysia, dinshah@ukm.my

Ratnam Wickneswari Faculty of Science and Technology, School of Environmental and Natural Resource Sciences, Universiti Kebangsaan Malaysia, 43600 Bangi, Selangor Darul Ehsan, Malaysia, wicki@ukm.my

Yoshihiko Tsumura Department of Forest Genetics, Forestry and Forest Products Research Institute, Tsukuba, Ibaraki 305-8687, Japan, ytsumu@ffpri.affrc.go.jp

K.M. Wong Singapore Botanic Gardens, Singapore 259569, wkm2000@gmail.com

Chapter 1
Forest Management Systems in Southeast Asia

Mohamad Ismail Shaharuddin

Abstract A number of systems have been developed and formulated for the management of tropical forests of Southeast Asian countries over the years. However, almost all countries have adopted some form of selective felling system to sustainably manage their forests. In the Philippines, selective logging systems have been adopted. Similarly in Indonesia, a selective cutting system was approved for implementation in 1972 which was revised and modified in 1989. The modified system is called selective felling and replanting system. In Malaysia the selective management system was introduced in 1978 replacing the successful Malayan Uniform System which was practised in lowland dipterocarp forests. The selective systems are generally designed for shorter cutting cycles ranging from 25 to 35 years. Under this system a pre-felling inventory is conducted to determine the cutting limit of trees to be felled. Immediately after harvesting a post-felling inventory is undertaken to determine the quality and composition of the residual stand followed by silvicultural treatment to enhance the residual stand as the future crop for the subsequent cycle.

Keywords Silviculture · Dipterocarp · Annual allowable cut · Forest inventory · Residual · Selective Management System · Tree marking · Growing stock

1.1 Introduction

The rapid socio-economic development in Southeast Asia particularly infrastructure, agriculture and industrial development has affected the level of timber production and other forest ecosystem services. At the end of 2010, it was estimated that the total forested area in Southeast Asia was 214 million ha which covers 49% of the total land area. The forest cover ranges from 26% in the Philippines to 68% in Laos PDR. In terms of forest cover loss there has been a decrease from 1.0% per annum in the 1990s to 0.3% during the period 2000–2005 and increase of 0.5% from 2005 to 2010 (FAO 2011).

M.I. Shaharuddin (✉)
Institute for Environment and Development (LESTARI), Universiti Kebangsaan Malaysia (UKM), 43600 UKM Bangi, Selangor Darul Ehsan, Malaysia
e-mail: dinshah@ukm.my

R. Wickneswari, C. Cannon (eds.), *Managing the Future of Southeast Asia's Valuable Tropical Rainforests*, Advances in Asian Human-Environmental Research 2,
DOI 10.1007/978-94-007-2175-3_1, © Springer Science+Business Media B.V. 2011

The forests of Southeast Asia have long been valued as a source of food, fuel, medicine and materials for shelter and livelihood. It will continue to play an important role in rural economies and influence the socio-economic development of the respective countries and finally the quality of life.

1.2 Forest Management System

Generally there are two types of management systems being practised in the Southeast Asian tropical rainforest via monocyclic or polycyclic system. The monocyclic system comprises of a uniform system, tropical shelterwood system, and irregular shelterwood system. The polycyclic system is mainly selective cutting of crop trees. In this regard, Troup (1928) described the selection system of silviculture as that where felling and regeneration is distributed continuously over the whole area in a forest which is wholly unevenly aged. The felling consists of the removal of a single tree or small groups of trees scattered throughout the forest.

According to Florence (1981, Silviculture II (D44) lecture notes, unpublished) the selection system in practice can be classified into two categories, viz. the 'narrow sense' and the 'broad sense'. The 'narrow sense' is typical of the classical selection system practised in Europe. The 'narrow sense' of selection system would apply where

- a balanced size class distribution is determined which will ensure full site stocking, good growth of trees of all size classes, and enough space for a regeneration class after logging;
- all size classes occur within each relatively small unit of forest;
- the growing stock is maintained at or near full site use, that is, logging is frequent and light; and
- there is regular and prescribed yield of the full range of products from each unit of forest.

In contrast, selective logging in the 'broad sense' refers to any form of selective cutting designed to meet specific objectives of management and not necessarily meeting any of the above criteria. Florence (1978) stressed, for example, that the silvicultural decision may be strongly influenced by the role of the forest in wood supply which can be biased either towards short-term or long-term commitments. Where maintaining supply for the short term is a problem, one obvious approach would be to conserve for later cutting as much growing stock as possible in merchantable or near-merchantable sizes whenever stand conditions permitted. This could involve selection-type logging and treatments to stimulate the diameter increment of immature trees without necessarily seeking to achieve a balanced size class distribution.

The so-called selection system adopted in many tropical countries differs very much from the classical selection system as practised in uneven-aged forests of

Europe and the United States. The selection system applied in tropical forests can be grouped largely under the 'broad sense' approach to selective logging. The selection approach in a number of Southeast Asian countries is reviewed.

1.3 Selective Logging in the Philippines

The selective logging adopted in the Philippines' dipterocarp forest is a modified form of the classical selection system (Virtucio and Torres 1978). The selective logging is described as 'the systematic removal of the mature, over mature and defective trees in such a manner as to leave an adequate number and volume of healthy residual trees of desired species necessary to assure a future crop of timber and forest cover for the protection and conservation of soil and water'. The silvicultural and management objectives of this system are

- to save as many as possible of the commercial tree species with future sawlog potential from logging damage during the harvesting of the merchantable timber from the virgin forest and
- to improve the residual growing stock through silvicultural treatments.

Reyes (1962) gave the following reasons why a selective logging system was adopted in the Philippines:

- It is the cheapest and surest measure for restocking the dipterocarp forest.
- Due to its stand structure, reproduction habits and the response of secondary storey trees to increased light, the dipterocarp forest can be appropriately managed under some form of selection system.
- The stocking of younger trees of many size classes may be adequate to restock the forest; these can be saved through careful logging and will survive and grow vigorously.
- The residual growing stock will reach sawlog size much sooner than newly established seedlings.

1.3.1 Selective Logging Procedures

As applied in the Philippines, selective logging has three basic phases.

1.3.1.1 Phase 1: Tree Marking

Tree marking is a prerequisite for selective logging, requiring the identification of an adequate number of trees to be left for the future crop. The residual crop which should include a component of young, healthy, well-formed commercial tree species is selected and marked. This is done after systematic sampling of growing stock at 5% intensity using 0.1 ha circular plots (Fox 1967). The goal of tree marking is to retain 70% of the stocking of trees within the 20–60 cm diameter class and 40% of the stocking of trees more than 60+ cm. Trees to be felled are marked with an arrow indicating the direction of felling.

1.3.1.2 Phase 2: Inventory of Residual Growing Stock

An inventory of growing stock is carried out after the logging operation and is designed to determine the effects of logging on the condition of trees marked for retention. These trees are then classified into healthy, substandard or damaged (Fox 1967). The healthy trees left serve as a basis for predicting the volume to be harvested in the next cutting cycle; the criteria for classifying residual trees are given in the *Handbook of Selective Logging* as follows (Glori 1979):

Healthy residuals

- A tree to be considered a healthy residual must be sound, thrifty, with straight cylindrical bole, uninjured or with slight injuries, for example, not more than 1/6 of the crown being severed or badly damaged;
- Any injury on the trunk reaching the wood (beneath the cambium) should not be more than 5 cm wide and 50 cm long, more or less, along a straight line parallel to the longitudinal axis of the trunk;
- Not more than ½ of the circumference at any place on the trunk should be girdled and deeply indented by wire rope;
- Not more than 1/6 of the root system should be removed or disturbed.

Substandard doubtful residuals

- A tree may be considered a doubtful residual when more than 1/6 but no more than 1/2 of the crown is severed or badly damaged;
- Any injury on the trunk over 5 cm wide and reaching the wood to affect no more than 1/3 of the circumference, and to be no more than 2 meters in length along a straight line parallel to the longitudinal axis of the trunk;
- More than 1/2 but no more than 2/3 of the circumference of the trunk may be girdled or seriously indented by wire rope;
- Over 1/3 but no more than 1/2 the number of buttresses should be severely damaged

1.3.1.3 Phase 3: Timber Stand Improvement

This phase includes all possible cultural operations from planting of landings with *Albizia falcataria* or other species to a liberation treatment carried out 5–10 years after logging. Such a treatment might include cutting of vines, deformed trees and unwanted vegetation and girdling of those trees which compete or interfere with the growth of desired tree species (Fox 1967, 1976).

Fox (1967) has been very critical of the theoretical basis of the selective logging system practised in the Philippines. There has been a tendency to implement the system over the whole forest area irrespective of its inherent characteristics or the way it varies with topography, terrain and species composition. Fox (1967) also argued that the success claimed for the system was based on a very small localized observation made in the Basilan forest in Mindanao.

The dipterocarp forests approach optimum development in the Philippines, and a good size class range is normally present in them compared to the forests in Peninsular Malaysia and Sabah where the intermediate size classes are poorly represented (Walton 1954). Against this background Nicholson (1970) believes that the dipterocarp forest in all the climatic regions of the Philippines can probably sustain a 30–45 year felling cycle provided the prescribed management is carried out. However, the careful selective cutting has seldom been undertaken by concession holders. Reyes (1978) firmly believes in the soundness and practicability of selective logging for the conservation and utilization of the dipterocarp forest, but the foresters and concession holders must make it work.

1.4 Indonesian Selective Cutting System

According to Wirakusumah (1977) the Indonesian Government approved in 1972 the following silvicultural systems for the management of the tropical rainforests of Indonesia:

1. Indonesian Selective Cutting System, also known as Tebang Pilih Indonesia (TPI)
2. Clear cutting with replanting
3. Clear cutting with natural regeneration

From this time, almost all timber concessions have been implementing the Indonesian Selective Cutting System.

1.4.1 Elements of the Indonesian Selective Cutting System

Basically the Indonesian Selective Cutting System involves the following activities:

1.4.1.1 Tree Marking

The marking is carried out to determine trees to be felled and also trees to be left as mother trees for the future crop. The number of mother trees, the diameter limit of felled trees and the cutting cycle can be determined from the guidelines presented in Table 1.1.

1.4.1.2 Inventory of Residual Stand

An inventory is carried out a year after logging to assess the nature of damage to the mother trees and the general condition of natural regeneration.

Table 1.1 Diameter limit, cutting cycle and the number of mother trees to be retained under the Indonesian Selection Cutting System

Diameter limit of felled trees (cm)	Cutting cycle of mother trees (years)	Number of trees/ha (cm)	Minimum diameter of mother trees	Annual allowable cut[a]
50	35	25	35	$1/35 \times M$
40	45	25	25	$1/45 \times M$
30	55	40	20	$1/55 \times M$

M = Volume of trees.
[a]Annual allowable cut for 35 years. Cutting cycle = $1/35 \times$ volume of trees.
For example, if the cutting limit of economic tree species is fixed at 50 cm dbh, then at least 25 mother trees of 35 cm dbh must be retained. The cutting cycle is 35 years. Hence by reducing the cutting limit to 30 cm dbh, more mother trees are required to be retained for the future crop.

1.4.1.3 Enrichment Planting Following Logging

Enrichment activities involve planting in areas where logging damage is severe. A nursery is established for this purpose to raise indigenous seedlings.

1.4.1.4 Cleaning and Weeding

These operations are carried out at two and five years after logging.

While almost all timber concession holders have implemented the Indonesian Selective Cutting System since 1972, the stipulated regeneration activities including enrichment planting have not been fully undertaken. According to Priasukmana and Montgomery (1979) less than 5% of the total number of concession holders undertook regeneration operations. They gave the following reasons:

- Maximum profit is sought by concession holders in the shortest possible time.
- There is uncertainty over the concession lease.
- There is insufficient government supervision of operations.
- The low financial rate of return.

In 1990, the Ministry of Forestry in their analysis on the situation and outlook of forest sector in Indonesia indicated that the implementation of the Indonesian Selective Cutting System encountered many difficulties in the forest areas. Many of the prescribed activities under the system were not carried out by the forest concessionaires (Yasman 1998). This has resulted in many inadequacies of the system with regard to the length of cutting cycle, growth rate of the residual species and the yield of the future timber crop.

Most observations seem to point to a gloomy future for the implementation of Indonesian Selective Cutting System. It must be emphasized that there has always been a tendency to apply the system on a blanket basis for all forests. The tropical rainforest varies considerably from place to place and between forest types. The system may need to be modified to suit individual forest conditions. Hence, the local forester has to utilize his technical knowledge prudently. Nicholson (1979) suggested that the Indonesian Selective Cutting System is basically sound but rarely implemented well in practice.

In 1989, a new system of forest management was introduced. The modified system is called Selective Felling and Replanting System (*Tebang Pilih dan Tanam Indonesia, TPTI*).

There are several basic principles to the *TPTI* system:

- stand inventory, including the structure of the regeneration, species composition and site characteristics;
- limits to the diameter and the number of trees to be cut in order to create a highly productive residual stand for the next rotation;
- maintenance of the residual stand, protecting the forest as a whole and maintaining the highly necessary biodiversity of the forest.

Tables 1.2, 1.3 and 1.4 show the scheduling summary of the activities required for the TPTI system.

1.5 Malaysia

Since the appointment of the first chief forest officer and the establishment of the Forestry Department in 1901, Malaysia (the then Malaya before gaining independence in 1957) has made marked progress in the area of forest management and silviculture. This is evident through the various management and silvicultural systems developed and refined over the years for the inland forests.

Table 1.2 Scheduling of the activities required for the TPTI system

Planning (years before or after harvesting)	Activity
−3	Design of the working area
−2	Pre-logging inventory
−1	Road engineering
0	Harvesting
+1	Refinement I
+1	Inventory of residual stand
+2	Production of planting stock
+2	Enrichment planting
+3	First tending
+4	Refinement II
+9	Thinning
+14	
+19	
Continuous	Protection and Research

Table 1.3 Scheduling of the activities required for the TJTI system

Planning (years before or after harvesting)	Activity
−1	Design of the working area
−1	Pre-logging inventory
−1	Road engineering
0	Harvesting
+1	Refinement
+1	Residual stand inventory
+1	Production of planting stock
+1	Planting
+2 to +5	Planting maintenance

Table 1.4 Scheduling of the activities required for the TTJ system

Planning (years before or after harvesting)	Activity
−2	Design of the working area
−1	Road engineering
−1	Production of planting stock
0	Harvesting of trees above 40 cm
0	Land preparation
0	Preparation of lines for plantation
0	Plantation
+1 till harvesting	Maintenance of plantation
Continuous	Plantation protection

1.5.1 The 'Gutta Percha' Era (1900–1910)

The first recorded forest management of importance in Peninsular Malaysia is generally known as the 'gutta percha' era (1900–1910) (Barnard 1954). Gutta percha is derived from the latex of various species of *Palaquium.* Many species produced such latex, but the best quality is known as getah taban merah and is derived from the trees of *Palaquium gutta.* The latex is white just like our rubber latex, but more viscous and coagulates to give a produce that is much less elastic. Gutta percha is one of the best known conductors of heat and electricity and will retain on cooling any shape given to it when hot. It is used in electrical, surgical and dental apparatus and in the manufacture of golf balls, but one of the most important uses was in the coating of submarine cables. However, the demand for gutta percha declined considerably with the introduction of radio telegraphy. As there were no proper technique, these trees were destructively felled for their latex which was priced as high as RM760/pikul (1 pikul = 60 kg) (Wyatt-Smith 1963a). Logging was highly selective and confined to the naturally durable heavy hardwoods, mainly *Neobalanocarpus heimii* (chengal) and *Shorea* species (balau) for use as railway sleepers. Hence, the early silvicultural operations were confined to the establishment of *P. gutta* and para rubber (*Hevea brasiliensis*) plantations and enrichment planting of mainly *N. heimii.*

1.5.2 Improvement and Regeneration Felling Era (1911–1942)

For the period 1911–1922, a series of silvicultural treatments known as *improvement fellings* were implemented, initially in Selangor and Negeri Sembilan, mainly to favour the development of *P. gutta* and poles of immature trees of the more valuable timber tree species. During this period, exploitation of timber was carried out exclusively on stateland forest with more than 90% of the wood consumed as firewood for domestic, mining, and rubber factories. A small proportion of the timber produced was utilized as railway sleepers. After harvesting, silvicultural treatments involved the removal of all species whose crowns were interfering with the valuable timber species and cutting of all bertam (*Eugeissona triste*) and climbers within a fixed distance (Wyatt-Smith 1963a). It was observed that natural regeneration was very good for all species except *P. gutta* and it was suggested that, instead of planting, it would be better to 'assist the existing natural regeneration by giving them more light and space'. These observations formed the basis of early forest improvement work which involved the release of young trees of economic value.

In 1922, a change occurred in the exploitation of the forests as forest reserves were opened for logging, mainly as there was an increase in demand for firewood and the mining boom had also created a strong demand for poles. These developments were considered in relation to the early improvement fellings, which were aimed solely at improving the existing forest stand. As the response of immature trees to these treatments was poor, doubt was expressed about the value of such treatments. Moreover, in the face of rising demand for timber, it was felt unwise to girdle trees which could be used as poles or firewood.

A review of the silvicultural practice in 1926 revealed that the value of many earlier improvement fellings which only favoured the growth of heavy hardwood species was not justifiable due to the rising demand for the medium and light hardwood species. However, it was also observed that young regeneration of the more valuable timber species had occurred, and the improvement fellings had in effect been regeneration fellings. As a result of this review, two classes of operations were distinguished:

Departmental regeneration improvement fellings (RIF): here the new natural regeneration arising from the fellings was assisted. It is generally confined to areas where there was no demand for inland firewood involving gradual or even drastic canopy opening, with two or more complete undergrowth cuttings. This operation was undertaken by the staff of the Forestry Department.

Commercial regeneration fellings (CRF): a development of the above system, carried out by private companies at a profit to the government for the production of poles and firewood. It involved the gradual removal, in several stages, of inferior species before the felling of the useful species was permitted by commercial operations, when there was a market for firewood and poles.

The departmental regeneration improvement fellings has no doubt proved itself, and some of the best regenerated forests in Peninsular Malaysia were the results of the application of the system. Ismail (1966) attributed the success of the system to the following factors:

Exploitation was carried out mainly over areas of rich lowland dipterocarp forest, easily accessible both for felling operations and for silvicultural treatment.

Ground preparation and light manipulation by clearing the understorey prior to any commercial felling created optimal conditions for seed germination and for the growth of the seedlings.

A series of selective felling operations carried out until the final crop was harvested, followed by a series of treatments in favour of the regrowth, further helped to establish the young crop.

Although the system involves a slow process of raising a new crop of even-aged trees, it is nevertheless a system under which the establishment of a new rich crop of chosen species can be assured. The system was possible and acceptable because the accessibility and species composition of the lowland dipterocarp forests made it quite economical to undertake preparatory treatments and a series of selective fellings.

Subsequently, with the advent of the Second World War and especially during the Japanese Occupation (1942–1945), extensive forest destruction occurred *to accessible forested lands*. During this period, uncontrolled exploitation of forest and felling of regenerated forests for food cultivation in both stateland forests and forest reserves without any silvicultural operations was carried out. Immediately after the war (1945–1948), forestry activities were centred mainly on assessing the status of exploitation in regenerated forest and research plots. By 1948, it was estimated that about 14,200 ha of forests under regeneration felling and 5,300 ha or nearly half of the fully regenerated forest available were destroyed (Wyatt-Smith 1963a).

1.5.3 Malayan Uniform System Era (1948 Onwards)

Following the Second World War, regeneration improvement fellings were discontinued because of the increased demand for raw materials by the increasing number of sawmills and by the use of heavy machinery in harvesting. Moreover, it was observed that many untended clear-fellings made during the Japanese Occupation contained adequate or even abundant regeneration of timber species and that it could, on the whole, look after itself. It was also observed that seedlings of *Shorea* spp. (Meranti) already on the ground responded immediately to sudden drastic canopy opening, thereby being able to grow well without much competition (Wyatt-Smith 1963b). These circumstances led to the formulation of the *Malayan Uniform System (MUS)* in 1948.

The MUS is basically a system for converting the virgin tropical lowland rainforest, which is a rich, complex multi-species and multi-aged forest, to a more or less even-aged forest, containing a greater proportion of the commercial species to be managed under the uniform system. It involved the felling, over a short single operation, of all economic tree species and immediately afterwards the poison-girdling of the uneconomic relic trees in order to release the seedling regeneration of the selected species. The adequacy of the seedling regeneration is determined by a systematic linear sampling known as the linear sampling milliacre (LSM). A summarized version of the sequence of operations under the MUS as originally prescribed is shown in Table 1.5 (Thang 1987).

The success of the MUS depended upon five main factors as follows:

- There must be an adequately well-distributed stocking of seedlings of economic species at the time of felling. This information is obtained from a milliacre (2 × 2 m) linear sampling of the area. If stocking is inadequate, the stand is to be left behind until after a seed year to be followed by another sampling.
- Complete removal of the canopy through poison-girdling of all remaining big-sized trees and other species considered as non-commercial down to 5 cm diameter at breast height (dbh).
- No tending until the regrowth had passed the ephemeral climber stage and was relatively clear below.
- Maintenance of adequate new canopy to prevent the redevelopment of climbers.
- Linear samplings to be carried out at regular intervals to assess the regeneration status of the seedlings.

The MUS was based upon experience gained in the lowland *Shorea* (Meranti)-rich forest, and the treatments prescribed were intended primarily to be applied to these forests under more or less ideal conditions. However, in many forests, there always appears to be an adequate amount of regeneration present. The stocking of regeneration, however, is not always uniformly distributed throughout the forest, and the condition of the forest may deteriorate after logging where a systematic assessment of regeneration had not been carried out and the timing of logging on a concession had not been coordinated with the presence of regeneration. Furthermore, where logging has been delayed for a number of years after a sampling which had shown adequate regeneration is present, that regeneration may

Table 1.5 Sequence of operations under the MUS

Year	Activities
$n - 1\frac{1}{2}$ to n	Linear sampling (milliacre) of regeneration and enumeration of commercial trees
n to $n + 1$	Exploitation, followed by poison-girdling of unwanted trees
$n + 3$ to $n + 5$	Linear sampling ($\frac{1}{4}$ chain square) (LS$\frac{1}{4}$) of the new crop followed by cleaning, climber cutting and poison-girdling as required
$n + 10$	Linear sampling ($\frac{1}{2}$ chain square) (LS$\frac{1}{2}$) of new crop, followed by treatment as required or passing as regenerated
$n + 20$, $n + 40$, etc.	Sampling and thinning as required

Table 1.6 Sequence of operations under the 'modified' MUS

Year	Activities
n–2 to n–1	Pre-felling forest inventory of 10% sampling intensity using systematic line plots. Analysis of data and determining volume and number of trees to be harvested (+45 cm dbh)
$n-\frac{1}{4}$	Tree marking of trees to be felled incorporating direction of felling
N	Felling of all commercial and utilizable species of 45 cm dbh and above
$n+\frac{1}{4}$ to $n+\frac{1}{2}$	Forest survey to determine fines on trees unfelled, royalty on short logs and tops and damage to residuals
n+2 to n+5	Post-felling forest inventory of 10% sampling intensity using systematic line plots to determine appropriate silvicultural treatments. Generally, poison-girdling of all defective and unwanted trees down to 15 cm dbh and climber cutting. Retention of advanced growth of potentially commercial trees
n+20	Forest inventory of regenerated forest to determine status of the forest

have died or the stocking reduced by the time of felling. Therefore, the availability of regeneration and the scheme of logging need careful co-ordination and planning under the MUS.

As the MUS focused mainly on the seedlings and saplings which were expected to form the bulk of the next crop, the silvicultural treatments were aimed at favouring these groups, often at the expense of bigger trees and advanced growth. This had led to a heavier poison-girdling of trees than was necessary and, in some cases, to drastic opening of the canopy. Hence over the years, the MUS was 'modified' with the emphasis shifted from seedlings and saplings to advanced growth. This has also resulted in a more discriminating use of the poison-girdling technique and a more conservational approach in silvicultural treatments. The sequence of operations under the 'modified' MUS is as shown in Table 1.6 (Thang 1987).

As a management system, the MUS had been considered successful in regenerating substantial areas of the lowland dipterocarp forests. In fact, many of the existing regenerated lowland dipterocarp forests found throughout Peninsular Malaysia would attest to that fact.

1.5.4 Selective Management System Era (1978 Onwards)

The MUS which was applied successfully in the lowland forest was found to be unsuccessful for management of the hill dipterocarp forest, the primary reasons being the comparatively more difficult terrain, uneven stocking, lack of natural regeneration on the forest floor before logging and uncertain seedling regeneration after logging due to irregular flowering of potential mother trees. In addition, the danger of erosion on steep slopes, the incidence of *E. triste* (bertam) and other secondary growth also did not favour a drastic opening of the canopy.

Moreover, virtually all easily accessible lowland and hill forests had been converted, particularly in the 1970s when the government embarked on a large-scale agricultural development scheme, such as FELDA and FELCRA, to eradicate rural poverty. Hence, the remaining virgin forests were found on steeper slopes in the remote parts of the country, mainly in the central/eastern region. Apart from the silvicultural and environmental difficulties encountered in trying to apply the MUS in the hill forests, the monocyclic approach was not financially, economically and ecologically attractive for implementation. The long rotation and the single cycle of the MUS would require a larger forest base for the practice of sustainable forest production. Since the availability of forest land is limited, it was envisaged that this could be achieved either by shortening the rotation or by increasing the number of cuts per rotation.

One of the major shifts in emphasis with respect to the management of the tropical forest in Peninsular Malaysia was the realization that the diversity of the flora and timber species need not necessarily be disadvantageous to the future wood production policies and strategies of the nation. On the contrary, a mixed forest crop offers the best cover for soil and water conservation and the wood-based industry should be able to utilize a wider spectrum of timber species in the future. In addition, technological innovations and sophistication in wood processing coupled with the changing pattern of wood usage in the forestry sector will render marketable a lot of presently non-commercial timber species in the future, in both the domestic and export markets.

Consequently, the selective management system (SMS) was introduced to allow for more flexible timber harvesting regimes which are consistent with the need to safeguard the environment and at the same time to take advantage of the demands of the timber market. More importantly, it discourages the poison-girdling of many of the presently non-commercial species which will conserve not only the wood but also the genetic resources available in the forests. The SMS was evolved to meet the following requirements:

- It has the flexibility to manage highly variable forest conditions and changes in the socio-economic environment.
- It is rationally based on inherent characteristics of the forest and prevailing socio-economic conditions.
- It will allow for the optimization of forest management goals which are, namely, (a) an economic cut, (b) the sustainability of the forest and (c) the minimum costs for forest development.

The implementation of SMS involves conducting forestry activities that could be distinctly categorized into three stages, namely pre-harvesting, during harvesting and post-harvesting activities. The pre-harvesting activities include pre-felling forest inventory, cutting limit prescription and timber tagging. During harvesting activities include directional felling and forest road construction while post-harvest activities include forest survey, post-felling forest inventory and prescription of silvicultural treatments. These activities are further elaborated below.

1.5.5 Pre-harvesting Activities

1.5.5.1 Pre-felling Forest Inventory

Prior to harvesting, a pre-felling forest inventory is normally conducted to provide reliable estimates of the population parameters being measured. The inventory includes sampling of all tree species of 5 cm dbh and greater, as well as tree species having dbh of less than 5 cm but having a minimum height of 15 cm. Results of the inventory include stand and stock tables (number of stems, basal area and gross and net volume on a per hectare basis) by dbh class and species groups. A stock map is prepared to indicate the distribution pattern of the tree size classes within the forest compartment.

1.5.5.2 Cutting Limit Prescription

Based on the stand and stock information obtained from the pre-felling forest inventory, together with other relevant information needed to determine the net economic production, the optimal cutting regimes (diameter limits) for the forest area were determined. In practice, the next cut under SMS is expected to be between 25 and 0 years, with an expected net economic out-turn of 30–40 m^3/ha, enriched with dipterocarp species. The criteria for cutting limit prescription are as follows:

- The cutting limit prescribed for the group of dipterocarp species should not be less than 50 cm dbh, except for *N. heimii* where the cutting limit prescribed should not be less than 60 cm dbh.
- The cutting limit prescribed for the group of non-dipterocarp species should not be less than 45 cm dbh.
- The residual stocking should have at least 32 sound commercial trees per hectare from the diameter class 30–45 cm or its equivalent.
- The difference in the cutting limit prescribed between the dipterocarp and that of the non-dipterocarp species should be at least 5 cm.
- The percentage of dipterocarp species in the residual stand for trees having 30 cm dbh and above should not be less than that in the original stand.

1.5.5.3 Timber Tagging

Subsequently, all trees above the prescribed cutting limits are marked (with direction) for felling. Timber tagging is carried out where harvestable trees are marked using plastic tags showing the licensee's identification code number, the serial number of the trees and the number of merchantable logs. This activity is carried out to ensure that only marked trees are felled, as well as to control the amount of timber output from the forest. Timber tagging has been computerized in some states and only logs that contain valid timber tags are allowed to be transported out of the forest. The verification process is carried out at the forest checking station at the time of royalty assessment. The timber tagging system has proven to be an efficient mechanism in controlling and tracking the movement and removal of logs from the forest.

1.5.6 During Harvesting Activities

During harvesting, prescribed forestry activities would have to be conducted in accordance with rules and regulations as stipulated in the logging license issued by the state forestry department. Among others, matters given due consideration during forest harvesting include the following:

- directional felling to ensure minimal damage to residual stand;
- construction of forest roads, skid trails and log landings according to prescribed standards to ensure minimal adverse environmental impact; and
- demarcation of adequate buffer zones along rivers and streams to mitigate soil erosion.

1.5.7 Post-harvesting Activities

1.5.7.1 Forest Survey

Immediately after harvesting, a forest survey is carried out to determine the number of tagged trees unfelled, short logs and tops left in the forest, as well as to assess damage on the residual stand. In the process, the amount of fines imposed due to activities not carried out in accordance with specifications is also determined. This information is included in the closing report for the license area.

1.5.7.2 Post-felling Forest Inventory

Normally, at 2–5 years after harvesting, a post-felling forest inventory is conducted to assess the status of the residual stand, as well as to determine any appropriate silvicultural treatments to be carried out.

1.5.7.3 Silvicultural Treatment

Based on the results of the post-felling forest inventory, the following types of treatment would generally be applied. They include girdling of defective trees (G), climber cutting (CL) or enrichment planting (EP) for areas that are poor in natural regeneration. A similar inventory is conducted at year 10 to assess the status of the regenerated forest. The sequence of operations under SMS is shown in Table 1.7.

The SMS, implemented since 1978 in Peninsular Malaysia, was recognized to be the most encouraging and complete theoretical system for operational forest management and provides for stand modelling based on adequate inventory and

Table 1.7 Sequence of operations under the SMS

Year	Activities
n–2 to *n*–1	Pre-felling forest inventory of 10% sampling intensity using systematic line plots to determine appropriate cutting regimes (limits)
n–1 to *n*	Tree marking incorporating directional felling
n	Felling all marked trees
n+1/4 to *n*+1/2	Forest survey to determine fines on trees unfelled, royalty on short logs and tops and damage to residuals
n+2 to *n*+5	Post-felling forest inventory of 10% inventory using systematic line plots to determine residual stocking and appropriate silvicultural treatments
n+10	Forest inventory of regenerated forest to determine status of the forest

data from permanent sample plots (Poore et al. 1989). Similar systems are being practised in the states of Sabah and Sarawak.

Besides the inland forests, the history of mangrove forest management dates back to 1903 when the first working plan was prepared by A.E. Wells (Gan 1995). Today, the mangrove forest located at the Matang Forest Reserve, Perak, is regarded as the best managed in the world. Since 1950, a total of six 10-year comprehensive working plans spanning the period 1950–2009 have been developed and successfully implemented for the mangrove forest at the Matang Forest Reserve. These plans laid out clearly the cutting cycle adopted which varies from 20 to 30 years, as well as a list of silvicultural treatments to be undertaken before and after the final harvesting.

1.5.8 Selection System in Sabah

Silviculture has been practised in Sabah since the 1930s. In the early days, silvicultural treatments were in the form of improvement fellings. When the Malayan Uniform System was formulated in Peninsular Malaysia, improvement feelings were abandoned and Sabah adopted a modified form of the Malayan Uniform System in 1956. This is fully described by Nicholson (1965). In 1971, as result of a study by Dr. H.C. Dawkins, the system was further modified (Chai and Udarbe 1977).

According to Liew (1978, The development of silvicultural policies for dipterocarp forest on Sabah, Unpublished MSc Thesis) the current silvicultural system adopted for the management of the permanent forest estate of dipterocarp forest in Sabah can be classified as a simple form of selective logging. The system emphasizes (a) saving the abundant seedlings already present in the regeneration pool and (b) retention of intermediate size class trees aimed at yielding an economic cut between rotations. Under the system, all trees greater than 1.8 m in girth are to be felled and removed. Intermediate size class trees from 0.3 to 1.8 m are to be retained as far as possible; the sequence of operations is summarized as follows:

Year	Activities
$F-2$	Allocation of coupe
$F-2$ to $F-1$	First silvicultural treatment, tree marking and climber cutting
F	Felling
$F+0$ to 1 month	Clearance inspection
$F+0$ to 2 months	First months assessment of regeneration based on LSM
$F+3$ to 6 months	Second silvicultural treatment – poison-girdling with 2,3,5-T in diesoline
F or $F=F$ or $F+10$ years	Establishment of yield plots
$F+5$ years	Second assessment of regeneration based on LS $\frac{1}{4}$
$F+5$ years	Third silvicultural treatment – poison-girdling. Treatments are prescribed according to the results of LS $\frac{1}{4}$. Should stocking be too low, for example, if 1/3 or less of the quadrats per acre are stocked with commercial poles or seedlings, enrichment planting may be carried out
$F+10$ to 15 years	Third year assessment of regeneration
$F+10$ to 15 years	Fourth silvicultural treatment – liberation and refinement

Liew (1978, The development of silvicultural policies for dipterocarp forest on Sabah, Unpublished MSc Thesis) stressed that the practicability of the selection system in Sabah is largely dependent on

- the effectiveness of climber cutting and tree marking prior to logging in minimizing logging damage;
- the abundance of intermediate size class trees;
- the response of the advance growth to canopy opening.

Although forest condition and composition varies considerably from place to place, Liew (1978, The development of silvicultural policies for dipterocarp forest on Sabah, Unpublished MSc Thesis) observed that the above conditions could be met in Sabah. Hence, a selective system could be applied in Sabah.

1.5.9 Selection System in Sarawak

Early silvicultural treatments in Sarawak were largely confined to peat swamp forest. Hence, logging in the mixed hill dipterocarp forest is a recent phenomenon. For this reason, silvicultural research in the hill forests of Sarawak has a relatively short history, starting on a significant scale only in 1970. Currently, silvicultural management is under active investigation with the assistance of the Food and Agricultural Organization (FAO) (Lee 1982).

In 1974, preliminary studies were carried out in a selected plot area called ‘guideline studies’. The results of the ‘guideline studies’ (Hutchinson 1977) showed the following:

(a) The residual stand contains a useful number of trees of desirable species which, with silvicultural treatment, are capable of producing a harvest on a short felling cycle.
(b) As long as selective harvests remain in moderate intensity, the mixed dipterocarp forest can be expected to maintain naturally the regeneration of sufficient stems of desirable species to warrant silvicultural treatment and management of the natural forest.
(c) The highest incidence of regeneration of desirable species occurs *where* an overwood canopy is retained after silvicultural treatment.
(d) Over a simulated time span of 20 years, most trees, except large ones with the poorest crowns, were seen to respond positively to release resulting from logging. Hence, it was concluded that they would respond the same way to silvicultural treatment. It was also observed that trees in the families of Dipterocarpaceae, Anacardiaceae, Burseraceae, Lauraceae, Leguminoseae, Sapotaceae and Sterculiaceae could be expected to respond positively to canopy opening.

From these results, it was concluded that the promise of a positive response to silvicultural treatment was sufficiently good to warrant experimental application of such treatments, and that the experiments should include types of treatment which maintain the better component of the residual stand (Hutchinson 1979). This led to the adoption of a post-logging 'liberation thinning' in the mixed dipterocarp forest of Sarawak.

The term 'liberation thinning' as applied in Sarawak refers to the process of selection of residual trees – as many as possible of the desirable species – and to liberate the best of them from competition with trees of lesser value (Hutchinson 1979). The frequency of silvicultural treatment suggested by Hutchinson (1980) is as follows:

Year	Activities
$F1 + 0$	Selective harvesting to a limit of 60 cm dbh
$F1 + 1$	Liberation thinning of residual forest and selecting final crop trees in the diameter range 10–59 cm dbh
$F1 + 30$	Second selective harvest
$F2 + 1$	Liberation thinning of residual forest (trees which were seedlings in open spaces created by the first harvest will be liberated at this time)

The development of silvicultural treatment in liberation thinning adopted in Sarawak is still at an early stage and was largely derived from plot studies. However, more intensive research is currently being carried out on a wider scale so as to obtain more conclusive results.

1.6 Conclusions

The choice of a silvicultural system depends on a host of factors which range from economic to ecological characteristics of a forest. Florence (1978) has identified and clearly elaborated the factors affecting the silvicultural decisions.

Economic considerations are by far the most important of the factors affecting the silvicultural decision in the tropical region. For example, the Shelterwood regeneration improvement felling system practised in the 1920s was ecologically sound and feasible, was replaced by the Malayan Uniform System in 1948 primarily for economic reasons. However, other factors may be just as critical in those forests, for example Florence (1978) has pointed out the need to cater for continuing short term log supply in the silvicultural planning in some tropical countries. The selective logging adopted in Peninsular Malaysia is aimed at conserving stock for future use.

Dawkins (1958) and Ashton (1978) have expressed scepticism about the success of the selection system in the tropical rainforest. Both advocate a uniform system as being more appropriate for the management of tropical rainforest. It must be stressed, however, that 'a good silvicultural system is not chosen but formulated as a solution to a specific set of circumstances' (Smith 1962), and that many relevant factors will vary from country to country and forest to forest as well. Thus it may be inappropriate to prescribe any one 'system' for the tropical rainforest.

Acknowledgement Comments from an anonymous reviewer are gratefully acknowledged.

References

Ashton PS (1978) The biological and ecological basis for the utilisation of dipterocarps. In: Eighth World Forestry Congress, Jakarta, Indonesia

Barnard RC (1954) A manual of Malaysian Silviculture for inland lowland forest. Research Pamphlet No. 14, Forest Department of Peninsular Malaysia, 199p

Chai DNP, Udarbe MP (1977) The effectiveness of current silvicultural practice in Sabah. Malays Forester 40(1):27–35

Dawkins HC (1958) The management of natural tropical high-forest with special references to Uganda, Imperial For Inst Paper No. 34

Florence RG (1978) The silvicultural decision. Forest Ecol Manage 1:293–306

Food and Agricultural Organization of the United Nations (FAO) (2011) Southeast Asia Subregional Report, Asia-Pacific Forestry Sector Outlook Study II: FAO Regional Office for Asia and the Pacific Bangkok, Thailand, 199p

Fox JED (1967) Selective logging in the Philippines Dipterocarp Forest. Malays Forester 30(3):182–190

Fox JED (1976) Constraints on the natural generation of tropical moist forest. Forest Ecol Manage 1:37–65

Gan BK (1995) A working plan for the Matang Mangrove Forest Reserve, Perak. Fourth revision State Forestry, Forest Department of Peninsular Malaysia, 214p

Glori J (1979) Silvicultural practices in the Philippines. In: Symposium on silvicultural technologies, Tsukuba, 1978. Tropical Agricultural Research Series No 12

Hutchinson ID (1977) Study to establish guidelines for silviculture and management of the mixed dipterocarp forests in Sarawak, Forest Department of Sarawak, 245p

Hutchinson ID (1979) Liberation thinning a tool in the management of mixed dipterocarps forest in Sarawak. In: Seventh Malaysian forestry conference, Penang
Hutchinson ID (1980) Approach adopted to define the interim guidelines for silviculture and management of mixed dipterocarp forest in Sarawak. In: International forestry seminar, Kuala Lumpur, Malaysia
Ismail HA (1966) A critical review of Malayan silviculture in the light of changing demand and form of timber utilization. Malays Forester 29(4):228–233
Lee HS (1982) The development of silvicultural systems in hill forests of Malaysia. Malay For 45(1):1–9
Nicholson DI (1965) A review of natural regeneration in the dipterocarp forest of Sabah. Malays Forester 28(1):4–26
Nicholson DI (1970) Forest management report FO: SF/PHI. 16 No. 3. FAO, Rome
Nicholson DI (1979) The effects of logging and treatment on the mixed dipterocarp forests of South East Asia. FO: Misc./79/8 Draft. Rome. 65
Poore D, Burgess P, Palmer J, Rietbergen S, Synnott T (1989) No timber without trees: Sustainability in the tropical forest. Earthscan Publications, London, 252p
Priasukmana S, Montgomery R (1979) Investment analysis of Indonesian selective cutting system in natural forest management. In: Seminar on Natural resources management in developing countries, Kuala Lumpur, Malaysia
Reyes MR (1962) Permanent forests – a must. Philippine Lumberman 21:22
Reyes MR (1978) Natural forests need attention. Canopy, December
Smith DM (1962) The practice of silviculture, 7th edn. Wiley, New York, NY
Thang HC (1987) Forest management system for tropical high forest with special reference to Peninsular Malaysia. For Ecol Manage 21:3–20
Troup RS (1928) Silvicultural system. Clarendon, Oxford
Virtucio FD, Torres MG (1978) Status of management and utilization of forest resources in the Philippines. Malays Forester 41(2):149–166
Walton AB (1954) The regeneration of dipterocarp forests after high lead logging. Empire Forestry Rev 33:238–344
Wyatt-Smith J (1963a) Intermediate felling in the Red Meranti forest. Malays Forester 26(4): 111–119
Wyatt-Smith J (1963b) Manual of Malayan silviculture for inland forest – Malayan forest records, No. 23. Vols. I and II. Forest Department of Peninsular Malaysia
Yasman I (1998) Improving silvicultural techniques for sustainable forest management in Indonesia 'Tropenbos seminar proceeding. Research in tropical rainforests: its challenges for the future' pp 205–211

Chapter 2
A Biogeographic History of Southeast Asian Rainforests

K.M. Wong

Abstract The species-rich ecological assemblages of SE Asia have been molded via both historical and ecological processes. Over both short and long timeframes, environmental inconstancy and heterogeneity play a fundamental part in shaping speciation and organismal diversity. Climate change has been important in past transformations of SE Asian ecosystems, which can bounce back, change, or go extinct. This account has brought together botanical and zoological examples, integrating palaeontological and archaeological information, as well as insights from molecular studies, to provide a synoptic look at the biogeography of SE Asia. The coming together of the sciences has enriched our understanding of the connections among landforms, ecosystems and species, the role of changes in the environment, both past and present, and how some losses brought about by anthropogenic activities could be minimized.

Keywords Biogeography · Climate and ecosystem change · Dispersal · Extinction · Island biogeography · Landscape evolution · Molecular phylogenetics · Vicariance

2.1 Biogeography: Organisms and Communities in Space and Time

Organisms grow, change in form, reproduce, and die. Many move around, dispersing in various ways. Plants, typically more sedentary, have their fruits, seeds, or other parts distributed by other organisms (such as birds or mammals), wind, water, and gravity, and so come to occupy different sites, areas, and ranges over time. Organisms sharing space contemporaneously form a community. Thus, communities are dynamic, changing in composition as members die out, others newly establish, and still others maintain their presence. Some communities are transient on site and are replaced by other communities in time; colonizing plants in the new community at a site are called pioneers and the communities following are described as successional. When general composition and other attributes such as

K.M. Wong (✉)
Singapore Botanic Gardens, Singapore 259569
e-mail: wkm2000@gmail.com

R. Wickneswari, C. Cannon (eds.), *Managing the Future of Southeast Asia's Valuable Tropical Rainforests*, Advances in Asian Human-Environmental Research 2,
DOI 10.1007/978-94-007-2175-3_2,

Fig. 2.1 Lowland evergreen rainforest with large tree crowns on more clayey soils (*lower part*) abruptly crossing to *kerangas* (tropical heath) forest with small crowns in Endau-Rompin national park, Peninsular Malaysia. Photo: K.M. Wong

structure remain more or less the same over a long period of time, with only components (such as which and where trees die and which patches contain younger or more mature trees) changing randomly within communities attain some kind of dynamic equilibrium and are said to exist in climax. Within a particular environmental regime, the highest form of vegetation that can develop is known as a climatic climax. Worldwide, similar vegetation climatic climaxes make up a formation. The different formation characteristics of wet tropical areas, such as lowland evergreen rainforest (Fig. 2.1), lower montane rainforest, and peat swamp forest (Fig. 2.2), are collectively and generally referred to as tropical rainforest.

For tropical rainforests, biogeography addresses both historical and ecological aspects of how and when these forests and their species came to occupy their present-day extents. Historical biogeography reaches back into geologic time, seeking to understand the past influence of landscape, climate, and sea-level changes. Ecological biogeography considers how prevailing conditions, especially abiotic factors, sustain characteristic communities and affect biological distributions. The interactions of organisms among themselves and with co-occurring organisms, such as in competition for resources, obviously affect community composition (species richness) and ecology at local scales.

2.2 Tropical Rainforests of SE Asia

Worldwide, tropical rainforests are found in three main areas: Amazonian South America, the so-called Indo-Malayan region of Southeast (SE) Asia, and tropical west and central Africa (Richards 1998). The Indo-Malayan rainforest

Fig. 2.2 Inside a forest stand dominated by the dipterocarp *Shorea albida* in the Badas peat swamp, Brunei. Photo: K.M. Wong

region (Whitmore 1984, 1989, 1998) includes the Malay Archipelago (Whitmore 1984) and extends northward, as pockets, into parts of Myanmar, Indochina, the Xishuangbanna area in southwest China, and south into northern Queensland in Australia and the southwest Pacific (Primack and Corlett 2005; Whitmore 1998). The ever-wet core of this region has been recognized as a distinct plant geographical region called Malesia and comprises the Malay Archipelago and the Malay Peninsula, stretching from Sumatra to Papua New Guinea and including the Philippines. Malesia is distinguished by hundreds of distinctive seed plant and fern genera that do not occur in surrounding regions and, conversely, similar suites that are absent from Malesia (van Steenis 1950). Geologically, Malesia includes two continental shelves, Sunda (including Sumatra, Java, the Malay Peninsula, and Borneo as the main land masses) and part of Sahul (containing Papua New Guinea to the east) (Tjia 1980), the intervening area of deeper seas and generally smaller islands between them, and the Philippine archipelago to the north.

Tropical rainforests occur where there is a humid climate with high rainfall that, on average, exceeds potential evaporation in all months. Mean annual temperatures are 20–28°C, fluctuating just 4–5°C throughout the year. Mean annual rainfalls of 2000–3000 mm or more give a Tropical Wet to Tropical Superwet regime with few or no dry months (those with less than 100 mm rainfall) (Richards 1998), although there are occasional severe droughts (Walsh and Newbery 1999). Geographical position and terrain affect local rainfall situations, with seasonal variations in SE Asia due to prevailing monsoon winds in the region (Goh 2005). Above 1000 m elevation, it is cooler by at least some 10°C, with a Tropical Montane climate.

Abnormalities in the climate can be due to typhoons or tropical storms (Wong 2005), El Niño/Southern Oscillation (ENSO) events that bring warm phases and occasional drought in SE Asia (Walsh and Newbery 1999) that last 1–2 years and recur every 2–10 years (Enfield 1992; Philander 1998), and the spread of fires into forests during dry spells that increase CO_2 loads and bring smoke pollution over the region (Heil and Goldammer 2001). Volcanic eruption occasionally has an impact, such as the 1991 Pinatubo eruption that issued aerosol clouds over tropical regions of the Earth for weeks (Stowe et al., 1992).

2.3 Rainforests in Space and Time

As the earth system changes over geologic time, climates and ecosystems, including vegetation formations, can change (Jackson 2006). When did flowering plants (which dominate present-day tropical forests) and dipterocarps in particular (dominant in much of the wet forests of Sundaland), the rainforests, and other communities, arrive, disperse, and flourish in SE Asia? How did they arise and spread? This is dependent on the geographical/land, climatic, and sea-level changes that have occurred since the formation of SE Asia that would have modified distribution of land area and organisms.

2.3.1 SE Asia's Jigsaw Landscape

In the scenario of continental drift, where large portions of the earth's crust move over geologic time, present-day SE Asia is composed of various terranes (continental blocks or fragments) akin to a "giant jigsaw puzzle" (Metcalfe 1998) (Fig. 2.3). Fossil, stratigraphic, and geophysical data indicate that most of these terranes originated since the early Paleozoic (ca. 400 Ma), rifting from north or northeast Gondwana, the major southern landmass at that time. Such drift over the next 300 million years brought together crustal fragments to form a proto-SE Asia as a southeastern appendage to Eurasia. Metcalfe (1998) summarizes continental configurations for the Paleozoic and Mesozoic eras and Hall (1998) for the Cenozoic era (after 65 Ma).

Using a 10-year analysis (1994–2004) of high-precision GPS velocity fields across SE Asia, Simons et al. (2007) indicated that Sundaland is a well-defined tectonic entity including Indochina, the Malay peninsula, the Sunda shelf, the southeastern part of Sumatra, the western and northern parts of Java, and the major part of Borneo. Although it had been described as a continental shield or craton, heat-flow and seismic studies indicate otherwise, and Sundaland has experienced much deformation, with concomitant mountain building and development of deep sedimentary basins (Hall 2009). The geotectonic plates that formed insular SE Asia had diverse origins. East Borneo and west Sulawesi rifted from Gondwana around 100 Ma. Palawan, Mindoro, and southwest Borneo were derived from the China–Indochina terrane about 80 Ma. Borneo and west Sulawesi, attached for some time,

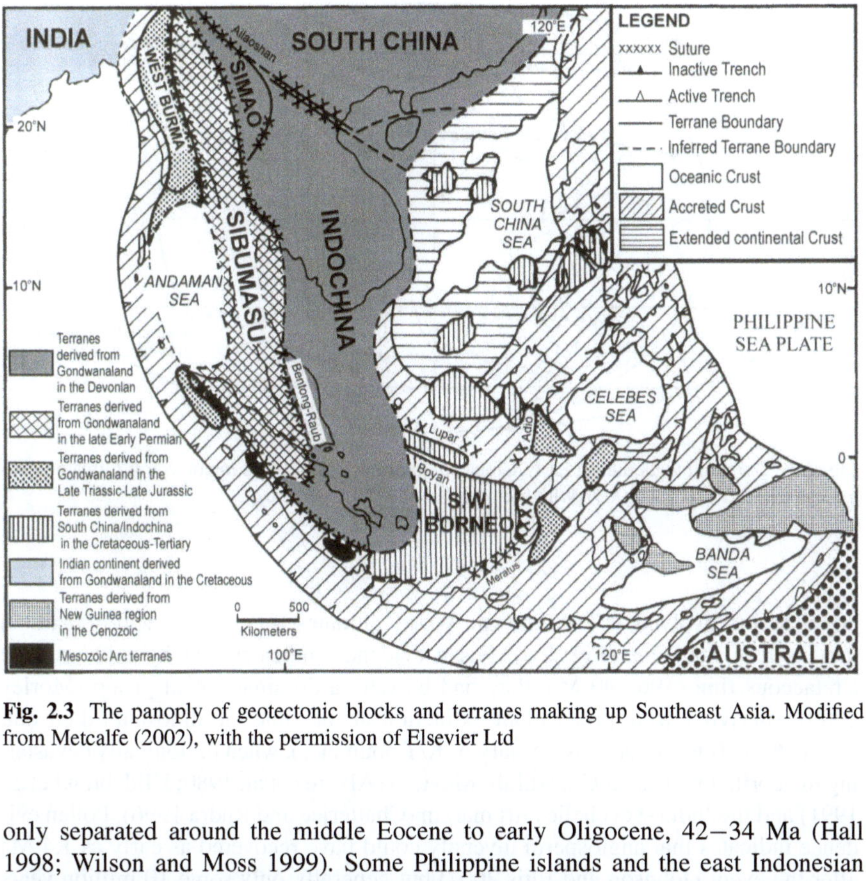

Fig. 2.3 The panoply of geotectonic blocks and terranes making up Southeast Asia. Modified from Metcalfe (2002), with the permission of Elsevier Ltd

only separated around the middle Eocene to early Oligocene, 42–34 Ma (Hall 1998; Wilson and Moss 1999). Some Philippine islands and the east Indonesian Halmahera Island formed a volcanic arc in the early Tertiary. The Barisan Mountains in Sumatra uplifted in mid-Miocene, 20–10 Ma (Kirby et al. 1993) or earlier, end Oligocene (Barber et al. 2005), and Java began as a series of volcanic islands coalesced from periodic volcanic deposits (end Pliocene, 5 Ma; Burrett et al. 1991). Australia's northward migration and "collision" with the Sunda plate from about 45 Ma (Hall 1998, 2009) not only renewed subduction at the south Sundaland margin, sparking widespread volcanic activity. It caused Borneo to rotate anticlockwise into its present position 20–10 Ma (mid-Miocene), and the uplift of the New Guinea mountains and the Banda Arc and other east Indonesian islands.

2.3.2 Biological Communities Over Geological Time

The geological history of tropical rainforests and their dominant plant group, the angiosperms (flowering plants), is now better understood thanks to intensive studies on the fossil record (Morley 2000). The earliest angiosperm fossils were found in early Cretaceous rocks (128–116 Ma) (Friis et al. 1994), although where exactly

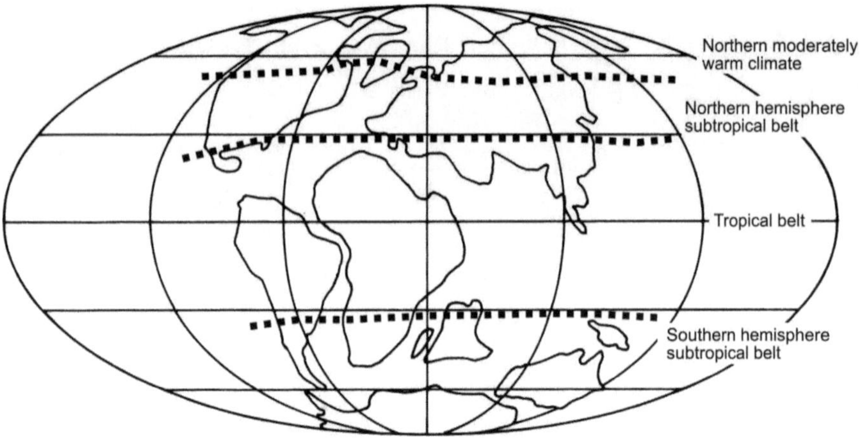

Fig. 2.4 The continents during early Cretaceous times, with Earth's major warm climate zones indicated. After Morley (2003), with the permission of Elsevier Ltd

angiosperms evolved is not known. Warm climates were more widespread on the Earth at that time (Fig. 2.4). Fossil evidence indicates that by mid- to Late Cretaceous times, 96–90 Ma, they had become a dominant plant group (Morley 2000), surviving widespread plant extinction at the time the dinosaurs died out, about 66 Ma (the Cretaceous/Tertiary or K/T boundary), when biologically devastating meteorite impacts hit Chicxulub (Mexico) (Alvarez et al. 1980; Hildebrand et al. 1991) and the India–Seychelles rift margin (Chatterjee and Rudra 1996). Pollen evidence indicates that angiosperm diversity could have recovered as early as 1.4 Ma after the event (Johnson and Ellis 2002) but generally only some 10 million years later (Morley 2000). Closed canopy tropical rainforests were significant over major parts of the Earth in the early Tertiary (54–49 Ma), including India and SE Asia (Morley 2000) (Fig. 2.5). At this time, the Indian plate, a fragment that had broken off from Madagascar and earlier from East Africa during Gondwana disintegration, collided against south Asia. Evergreen tropical rainforest elements carried on the Indian plate that crossed into SE Asia, as evidenced from pollen frequencies on both sides and other fossil and biogeographical data, included *Durio* (Malvaceae) and early Dipterocarpaceae (the "dipterocarps," western Malesia's dominant big tree forest family today) (Morley 1998).

In the Oligocene (ca. 35–25 Ma), general cooling and drying out saw contraction of rainforests worldwide, to just south America, Africa, parts of India, northern SE Asia, and east Australia (Harley and Morley 1995; Morley 2000). Throughout the Oligocene–Miocene, low-diversity dipterocarp monsoon forests were probably widespread in SE Asia, together with dry pine forests in Sumatra and Kalimantan and *Barringtonia–Lagerstroemia*-type swamp associations in basins (Muller 1981; Morley 1991). North temperate-type conifers (fossil pollen resembling *Abies*, *Picea*, *Tsuga*, etc.) were common in the more northern lowlands,

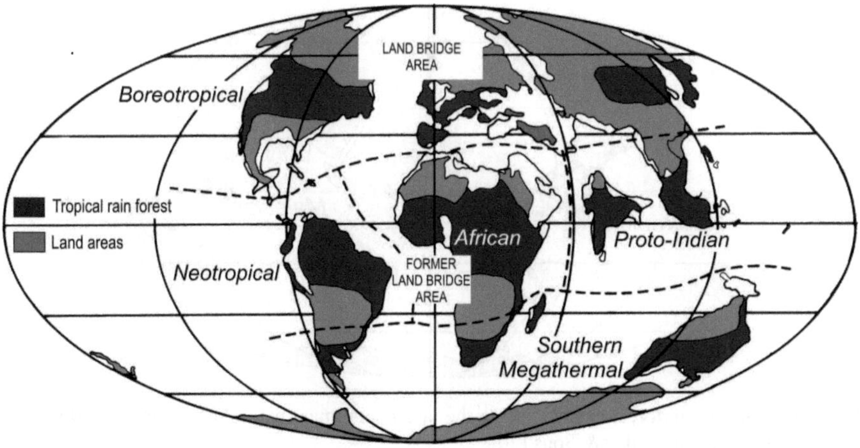

Fig. 2.5 Early Eocene reconstruction of landmasses and tropical rainforests. After Morley (2003), with the permission of Elsevier Ltd

together with temperate angiosperms over much of Peninsular Malaysia and northwest Borneo (Muller 1966). Some montane plant communities had been able to migrate to lower elevations (Morley 2000).

After the Early Miocene (ca. 20 Ma), warmer climates returned more extensive cover of tropical lowland rainforests to SE Asia. It was at this time that a diverse and distinctively "Malesian" flora (including dipterocarp taxa related to *Anisoptera*, *Dipterocarpus*, *Hopea*, or *Shorea*) developed (Morley 2000). Up to the Early Pliocene (ca. 5–3.5 Ma), distinctive communities could be discerned from pollen profiles, such as peat swamp forests, freshwater or streamside communities, and heath forests (Muller 1972; Morley 2000), and savanna vegetation (as profiles dominated by grass pollen) replaced some rainforest areas only intermittently in the Malay basin, northwest Borneo, the Mahakam delta, and Java during times of significant cooling, dryness, and lower sea levels.

Climatic and sea-level changes and corresponding fluctuations in SE Asian vegetation cover continued through the Quaternary, from ca. 1.8 Ma (Flenley 1979a, 1979b, 1984; Morley and Flenley 1987; Stuijts et al. 1988) (Fig. 2.6). In central New Guinea, rainforest contracted to as much as 75% of its present extent at the Last Glacial Maximum (LGM) (Walker and Chen 1987) and montane forest could have migrated as much as 1500 m downward (Walker and Flenley 1979). Lowland forest areas would have been colonized by cool-adapted vegetation but rainforests could also have migrated down to continental shelves exposed by sea-level reductions (Morley 2000). At various episodes, sea levels fell 40–120 m below present level and up to 2 million km^2 of the present submarine continental shelf would have been exposed, with dryland connections between the major areas of Sundaland (Heaney 1991; Voris 2000). Today, species found on two island groups in the South China Sea, the Anambas and the Natunas, may reflect such former land connections, including the gecko *Ptychozoon lionotum*, monitor lizard *Varanus nebulosus*, and

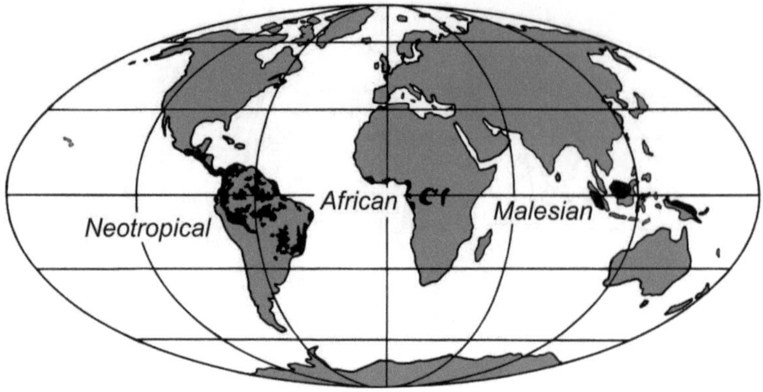

Fig. 2.6 Tropical rainforest extents during the last glacial maximum. After Morley (2000), with the permission of John Wiley & Sons Ltd

snake *Trimeresurus puniceus* (Malay Peninsula, eastern limit at the Natunas), and some Bornean species reaching westward only as far as these islands (Leong et al. 2003). A mid-Pleistocene comet collision at 0.793 Ma (Lee and Wei 2000) somewhere in Indochina incinerated populations and vegetation over a wide area there (Bunopas et al. 1999). The Indian subcontinent fell under more seasonal climates since the Late Miocene–Pliocene and rainforest there declined in extent, with much extinction of rainforest elements that formerly flourished, including dipterocarps (Harley and Morley 1995; Ashton 1982; Ashton and Gunatilleke 1987). In contrast, the dipterocarps became more diverse in SE Asia during the Oligocene–Pleistocene (Morley 2000).

2.3.3 A Savanna Corridor in Sundaland

The Pliocene–Pleistocene migration of mammals such as mastodons, stegodons, elephants, oxen, antelopes, and hippopotami, which are characteristic of open woodlands more common in seasonal parts farther north, to Java, where they are represented by fossils not found earlier than 2.4 Ma, could only have been possible via a suitable land route. The enabling route is envisaged as a savanna corridor that stretched from mainland SE Asia through the Malay Peninsula into the southern fringes of Sundaland (Medway 1972; Heaney 1991; Whitten et al. 1996; Cranbrook 2000; van den Bergh et al. 2001).

Past existence of a mid-Pleistocene pine savanna has been inferred from an abundance of pine (36%), grass (19%), and Compositae (2%) pollen in rocks near Kuala Lumpur (Morley and Flenley 1987; Morley 1998). Studies of stable carbon isotope composition of ancient cave guano profiles (Wurster et al. 2010) infer recession of rainforest limits from the Malay Peninsula and Palawan area, but not over northwest, north, and east Borneo, over the Last Glacial Period. Similarly dry-adapted

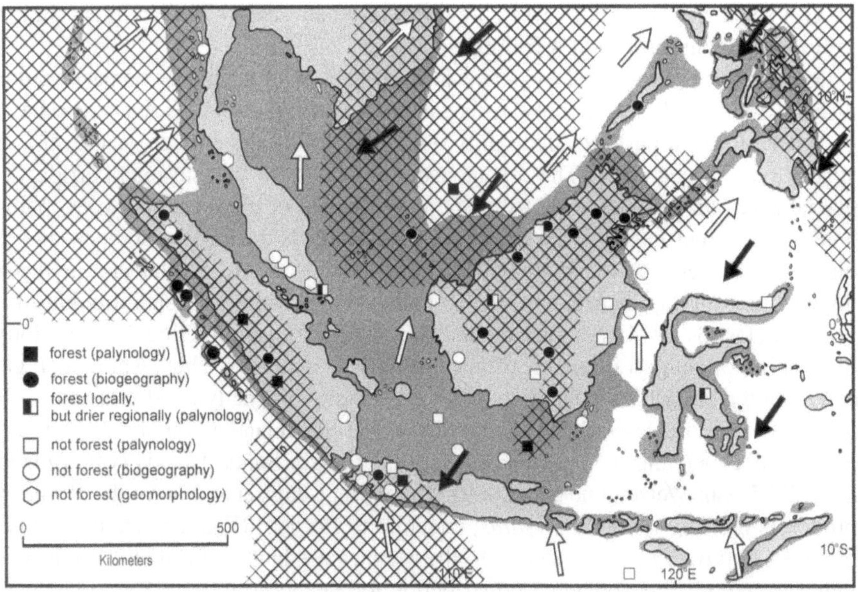

Fig. 2.7 Southeast Asia, showing maximum extent of emerged land area (*shaded*), locations of forested (*solid symbols*) and non-forested sites (*open symbols*) during the Last Glacial Period (Late Pleistocene), deduced from palynology (*squares*), biogeography (*circles*), or geomorphology (*polygons*). After Bird et al. (2005) with *added solid circle* for Sarawak indicating high inter-riverain endemism for dipterocarps (Ashton 1972). Winter monsoon (*black arrows*) and summer monsoon winds (*white arrows*) indicated as in Heaney (1991); *hatched areas* are those likely to have remained continuously wet over the Pleistocene according to the cumulative evidence depicted, with *unhatched areas* representing corridors of more seasonal climate, including the savanna corridor depicted in Heaney (1991). Adaptations from Heaney (1991) and Bird et al. (2005) made with the permission of Springer Science + Business Media and Elsevier Ltd., respectively

grassland or vegetation has also been inferred farther south and southeast and in mainland SE Asia (Morley 2000; Bird et al. 2005) (Fig. 2.7). Rain shadow effects of high mountains in Java and Sumatra at the LGM and differences in the relative strengths of the East Asian and Australasian monsoons could have resulted in reduced wetness in the center of Sundaland at such times (Bird et al. 2005). Geomorphological evidence such as Late Pliocene and early Pleistocene valleys of thick boulder beds and braided river and alluvial fan sediments (Verstappen 1997) and streamflow sediments found deeper than 200 m below the present sea also indicates Quaternary periods with more seasonal climate and reduced sea levels (Batchelor 1979). Seabed structural characteristics and eustatic processes are consistent with intermittent periods of land contiguity from south Malay Peninsula, through the Bangka and Karimata Islands, to south Sumatra, Java, and Borneo (Emery et al. 1972; Aleva et al. 1973; Lambeck and Chappell 2001; Bird et al. 2006). Such connections would have brought the agile gibbon, the three-colored

Prevost's squirrel, and ruddy tree shrew from Sumatra across the Karimata Straits to Borneo (Cranbrook and Piper 2008).

Modern biogeographical evidence is consistent with such a savanna corridor through Sundaland. Several present-day disjunct distributions in SE Asia could have been derived from past area contiguity during the LGM and climate, sea-level and vegetation changes since. The dry monsoonal climate of this savanna corridor in glacial times, in allowing establishment and dispersal of plant taxa between mainland SE Asia and Java and the Lesser Sunda Islands, is the basis of today's shared plant species between these areas that are absent elsewhere in Malesia, giving a disjunct present-day distribution, the so-called Asian/South Malesian drought disjunction (Jacobs 1974; van Welzen and Slik 2009). Prevailing climatic conditions play an important part in the persistence of floristic elements. Thus, while geologically being a part of Sundaland, much of Java, being distinctly more seasonal, still hosts a flora that is less related to that of everwet Sundaland and more characteristic of Wallacea (central Malesia), especially the lesser Sunda Islands (van Welzen et al. 2005; van Welzen and Slik 2009). The present-day separation of the Bornean proboscis monkey (*Nasalis larvatus*, endemic to Borneo) (Fig. 2.8) from its close relative (*Simias concolor*, endemic to the Mentawai Islands off Sumatra) probably indicates a more widespread past distribution for an ancestral form (Brandon-Jones 1996; Harrison et al. 2006). Today's large grazing or browsing herbivores could represent relictual populations of a past episode of savanna woodlands and more seasonal climates, such as the tapir *Tapirus indicus*, with fossils in Pleistocene Sarawak and Sabah but now restricted to the Malay Peninsula and Sumatra (Medway 1960; Cranbrook and Piper 2009).

Fig. 2.8 *Nasalis larvatus*, the proboscis monkey endemic to Borneo: male with harem. Photo courtesy of Arthur Y.C. Chung

2.3.4 Refugia for Tropical Rainforest Species

As tropical and subtropical vegetation began to fluctuate in extent over the Pliocene–Pleistocene, some peripheral areas in Sundaland appear to have retained somewhat continuously moist forest cover, in which forest species could conceivably "tide over" drier periods. Likely Pleistocene refugia on mainland SE Asia include SW Yunnan, indicated by near-tropical temperatures and greater winter humidity and rainfall that had existed then (Liu et al. 1986), and an apparently stable environment without significant faunal change (Ferguson 1993). The Burmese Mountains area and the Annamite Mountains east of the Mekong River probably had similar conditions and were likely ice-age refugia (Jablonski and Whitford 1999; Meijaard and Groves 2006). The many mammal species and endemics found in Laos and Vietnam compared to regions farther west in SE Asia support the idea of an Indochinese refugium (Meijaard and Groves 2006).

Similarly, forest cover appears to have been continuous through ice-age periods in west-central Sumatra, west Java, and on the Sunda shelf off northwest Borneo; this is suggested by palynological profiles (Stuijts et al. 1988; Maloney and McCormac 1995; Sun et al. 2000; Morley 2000) (Fig. 2.7). Palynology also indicates that a south Sulawesi site could have had forest patches in generally open vegetation (Hope 2001). Present-day soil-feeding termites, which would have required prolonged stability to build up diversity, indicate long persisting forested conditions in north and east Borneo, north and west Sumatra, and the Mentawai Islands (Gathorne-Hardy et al. 2002). These are along the fringes of Sundaland, where wetter conditions due to proximity to seas could have persisted. In the last glaciation, Sunda dry periods could have mainly affected the core regions, whereas peripheral areas could have been moderated by the east Asian winter monsoon that picked up moisture as it crossed the sea (Li and Sun 1999). The Australasian summer monsoon could have been drier overall (Wang 1999; Sun and Wang 2005) but still picked up moisture over the paleo-landscape before blowing over Java and Sumatra (Fig. 2.7).

A northwest Borneo refugium, still on the edge of the Sunda shelf just below the southern margin of the Pleistocene South China Sea, is especially evident. Ashton (1972) pointed out high endemism levels, especially among dipterocarps, in northwest Borneo and Corner (1960) had proposed that the "Riau pocket" represented phytogeographical similarities and impressive levels of local endemicity among presently separated sites in southeast Malaya and southwest Sarawak. These implied accumulated species richness in longer lasting stable environments in refugia. The area between Mt Kinabalu and Mt Dulit in northwest Borneo, where the lesser gymnure (*Hylomys suillus*) and ferret-badger (*Melogale everetti*) are restricted, may be part of this refugium (Brandon-Jones 2001). Cranbrook (2010) summarizes records that indicate a continued presence of arboreal primates throughout an archaeological profile at Sarawak's Niah Caves, suggesting a persistence of forest cover. The phylogeography of stoneoaks (*Lithocarpus*, Fagaceae) indicates that, in Borneo, there has been little migration and no detectable local extinction since the Late Eocene (Cannon and Manos 2003), so that widespread populations had little significant disruption even through the Quaternary ice ages. Sun et al. (2000) studied pollen assemblages from a South China Sea site off northwest Borneo, close to the discharge of the former North Sunda River, which was fed by tributaries from east Sumatra and west Borneo. At the LGM (25.4–19.7 ka) and until its end (19.7–10.4 ka), there was frequent alternation between dominance of montane rainforest gymnosperms and lowland rainforest taxa. Such lowland forest material in the profile could have been derived from surrounding low sites, but could also have been from refugial gallery forests along streams. There were 114 tropical lowland rainforest taxa and montane rainforest gymnosperms (such as *Podocarpus*, *Dacrycarpus*, *Dacrydium*, and *Phyllocladus*).

2.4 Dispersal, Vicariance, and Extinction: Organisms Past and Present

The juxtaposition of the Indian plate against south Asia during the Paleocene–Eocene permitted biotic interchange between the two terranes, including the arrival of dipterocarps in SE Asia. Thus, the earliest tropical rainforests with dipterocarps could only have existed in SE Asia from the Miocene (about 20 Ma), after arrival and further speciation of dipterocarps in moist forest conditions. The impacts of Quaternary ice ages, when much of the Sunda core dried out and was only suitable for savanna-type communities, make it likely that over much of the Malay Peninsula and south Borneo, continuous survival of dipterocarp rainforests into the present time would only have been after the retreat of the LGM from around 20 ka.

The overall richness of SE Asian rainforests seen today may derive from a number of reasons. One possible contributing factor was the extensive continental shelf or shallow sea area (largest in the tropics) and archipelagic character of much of the region. As sea levels changed between glacial periods, this sizeable shelf area accommodated the appearance and submersion of exposed land and the corresponding vertical migration of ecosystems, whether a cool-adapted savanna-type or a tropical moist vegetation. Thus, entire communities were probably not extinguished, especially where (in relict habitats functioning as refugia) local conditions remained suitable throughout the periods that profoundly altered the core Sunda areas. In cooler times with lowered sea levels, some rainforest organisms were able to migrate downward onto exposed continental shelves and some others probably survived in gallery forests along ancient rivers. The fluctuations in climatic and vegetation character, superimposed on colonization of newly available habitat or niches, probably provided more stress and selective pressure and acted as a "speciation pump."

The biogeography of SE Asian rainforests concerns the evolutionary and ecological history of its organisms and communities and the factors that have influenced the spread of their genetic signatures over components of the changing landscape. This is the interaction of rainforest organisms and their environment in space and time. The huge Sunda peninsula linking Indochina to insular Sundaland during drier climatic episodes facilitated latitudinal migration (dispersal) of terrestrial communities, which intruded into rainforest-occupied areas and replaced earlier communities. The peninsula acts as a barrier to the population (genetic) continuity of some marine organisms. As sea levels rose and the neck region of the Malay Peninsula became narrower and when climatic differences north and south of this became accentuated, many terrestrial populations encountered a gene-flow "bottleneck." Populations separated long enough by newly emerging barriers, such as intervening unsuitable habitat or wide enough sea stretches (as among islands), would have had the opportunity to diverge. Yet, whole groups of organisms may never cross some spaces because enabling corridors or dispersals never materialized. There are, of course, "turnovers" in populations and communities, as changing conditions may extinguish most or all organisms in some places.

2.4.1 Sunda Corridor

Some groups of SE Asian organisms that have been analyzed for their distribution illustrate quite well how their evolutionary development has been linked to episodes of geological history and past climate change. The phylogenetic–biogeographical study by Ridder-Numan (1998) shows insular SE Asian lineages of the leguminous liana genus *Spatholobus* and other plant genera distributed in ways that are consistent with the evolution of the region. The most recognizable clusters (clades) implying close relationship among members produced a nested hierarchy which, when area relationships were overlain and considered, implied that insular SE Asian taxa were derived but deeply diverged from mainland SE Asian ancestors. The Malay Peninsula acted as a migration corridor (when low sea levels gave maximal land and population continuity) and also gene-flow "bottleneck" (when high sea levels increased isolation of mainland SE Asian populations and taxa from those farther south of the isthmus). Early-diverging insular lineages were found in Sumatra, Borneo, Palawan, and Sulawesi, indicating dispersal over a wider Sunda landscape, again enabled during episodes of low sea level. A distinct lineage of species developed over different parts of Borneo (beginning in the Meratus and SW Borneo areas that were geologically older), inferring that after initial dispersal, relative isolation and local radiation produced the more derived taxa on the island.

A phenetic analysis by Yap (2002) of some 3500 species of East and SE Asian freshwater (cyprinid) fish also shows the close relationships among the Thai–Burmese, Indochinese, and Sundaland regions. For the common genera, SE Thailand and SW Cambodia were especially related to the Malay Peninsula, Sumatra, Java, and Borneo; but overall, a particularly close relationship between Sumatra and west Borneo, reflecting past contiguity, also existed. Inger and Voris (2001) point out that among 16 frog species that the Malay Peninsula shared with the Myanmar–Indochina area (mainland SE Asia), 6 also occur in Sumatra but only 1 in Borneo, probably due to the extent and duration of uninterrupted land contiguity. The phylogeographic study by Azuma et al. (2006) shows that the Asian weaver ant, *Oecophylla smaragdina*, has not only a coherent genetic haplotype (unique allelic

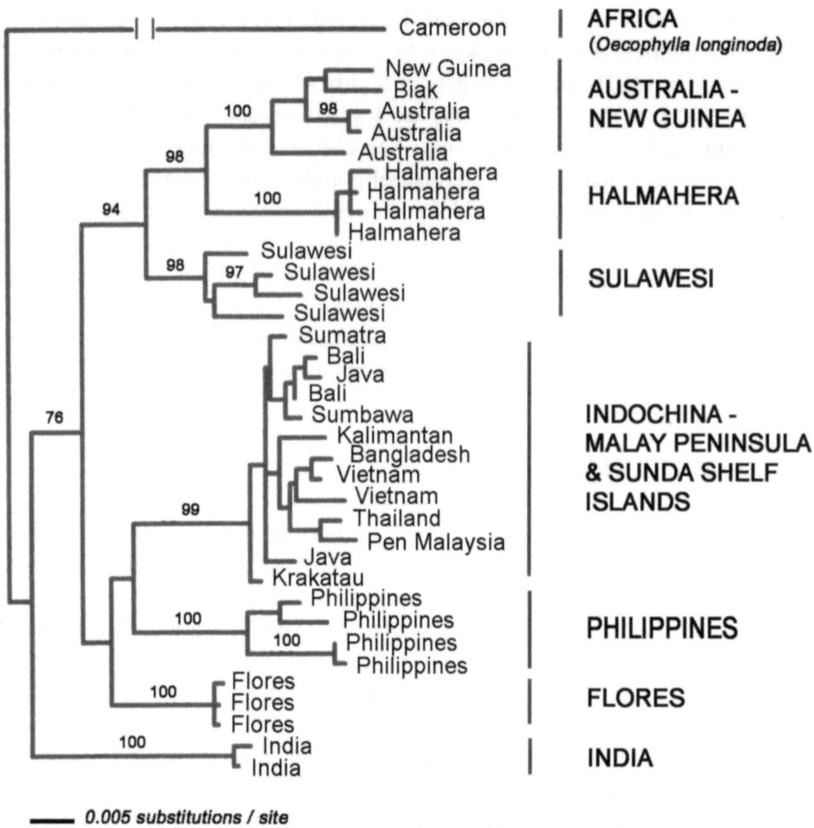

Fig. 2.9 Area relationships resulting from a phylogeographic analysis of the SE Asian weaver ant (*Oecophylla smaragdina*). Redrawn after Azuma et al. (2006), with the permission of The Ecological Society of Japan, Springer Japan, and the authors

pattern or DNA sequence variation) from northeast India–Indochina to the Malay Peninsula and Greater Sunda Islands, which began radiating around 1.6 Ma (early Pleistocene), but also distinct haplotypes for the Philippines, Sulawesi, Halmahera, and New Guinea (Fig. 2.9). This indicates sustained population exchanges (as would be enabled by land contiguity) in the past among the major Sunda shelf islands, and how distance and relative isolation from the distributional center could result in genetic divergence.

Macaca fascicularis monkeys range throughout Myanmar–Indochina, the Malay Peninsula, insular Sundaland, and the Philippines and have hybridized with the related *Macaca mulatta* that occurs from north India through most of China (Tosi et al. 2002). This hybridization is reflected in the incongruence between maternally inherited mitochondrial DNA and paternally inherited Y chromosome DNA phylogenies, due to the introgression of the *M. mulatta* Y chromosome into *M. fascicularis* populations (this is explained by the activity of male macaques

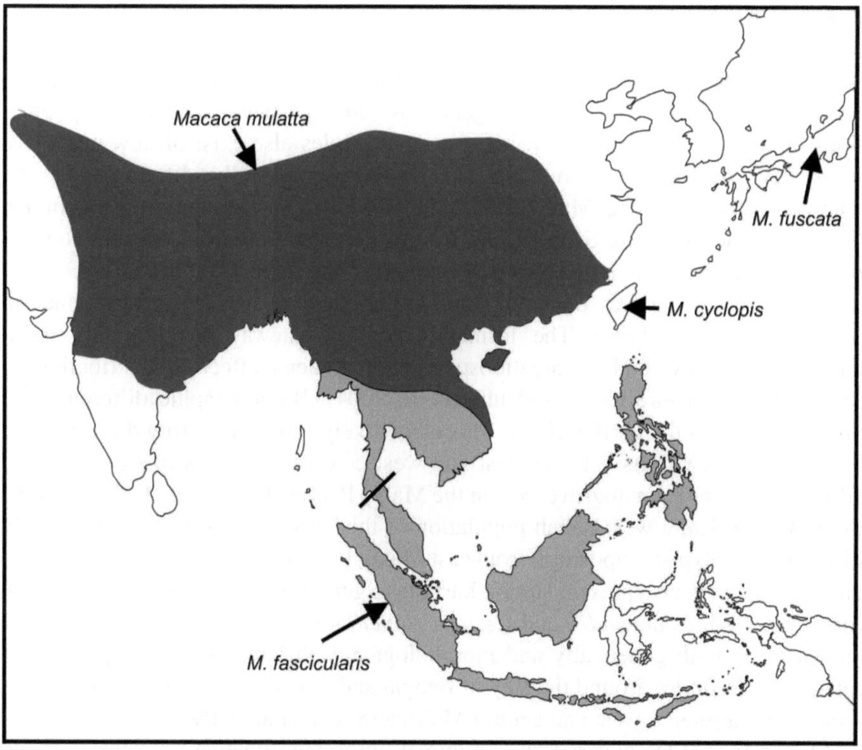

Fig. 2.10 Ranges of *Macaca mulatta* and *M. fascicularis* and related Asian species, with the southern limit of *M. mulatta* introgression indicated at the neck of the Malay Peninsula (*slash*). After Tosi et al. (2002), with the permission of Springer

that are not as territory bound as females, although reverse introgression has not been found). However, such introgression has not been detected south of the Isthmus of Kra (Fig. 2.10), where the Y chromosomes of male *M. fascicularis* are distinguished by a base deletion and a base insertion. This suggests that the Kra Isthmus represents reduced opportunities or unsuitable environment for transgression of *M. mulatta* males farther south, where conditions are wetter.

Thus, ecological barriers can exist in spite of land continuity. Frogs dependent on clear rocky streams on higher terrain for reproduction and fossorial-secretive snake species (living in forest litter or burrows), for example, would not easily cross drier, low-lying stretches of land, such as during periods of historical rainforest contraction (Inger and Voris 2001). This probably accounts for the lower similarity found for such frogs between the Malay Peninsula, Sumatra, and Borneo, compared to fundamentally lowland taxa that breed in ponds or slow waters; or the restriction of over 80% of fossorial-secretive snakes to just a single landmass today, compared with some 30–40% endemicity among terrestrial, arboreal, or aquatic snakes, which can disperse over vegetation or follow estuarine–coastal waters, as past land–sea distribution changed (Inger and Voris 2001).

2.4.2 Bottlenecks and Barriers

Apart from the examples given above, of dissimilar or unsuitable intervening habitat conditions preventing or reducing gene flow among spatially separated populations in spite of land continuity, interesting examples also exist of how mangrove forest and marine organisms may be phylogeographically differentiated. An emerged Sundaland and Malay Peninsula presents a barrier between the Indian Ocean and South China Sea. The impediment to free-flow inter-oceanic connectivity between the Indian and SW Pacific regions has been called the Indonesian or Southeast Asian Gateway or seaway, and the marine passages form the Indonesian throughflow (Hall 2009). The formation of this gateway has not only influenced throughflow and ocean/atmospheric phenomena affecting distribution of temperature–moisture regimes (Kuhnt et al. 2004). Biogeographic differentiation also ensues from the barrier effect. This is the likely cause of sharply differentiated chloroplast DNA traits between east and west coast populations of *Ceriops tagal* (Rhizophoraceae), a mangrove tree, in the Malay Peninsula (Liao et al. 2007). In the South China Sea, a west Sabah population of this species also shows significantly different haplotype composition from east Malay Peninsula populations, so distance and currents may also present similar barriers to genetic exchange (Liao et al. 2007). Similarly, *Ceriops decandra* and *Ceriops zippeliana*, confused in the past but now shown to be both genetically and morphologically differentiated, occupy distinct ranges: *C. decandra* around the Bay of Bengal and Andaman Sea and *C. zippeliana* along sea margins of west and central Malesia (Sheue et al. 2009).

Mitochondrial DNA variation is minimal, implying effective gene flow, among Sundaland populations of the coastal or mangrove-dwelling dog-faced water snake *Cerberus rynchops* (Karns et al. 2000). Still, there is significant divergence in genetic traits between Indian Ocean and South China Sea populations of the same species (Alfaro et al. 2004), indicating that the aggregation of Sunda landmasses is a gene-flow bottleneck between these oceans. The Kuda seahorse *Hippocampus kuda* also shows genetic differentiation between the Indian and Pacific oceans (Lourie et al. 2005); in this case, the Malesian continental shelf area would be the barrier or bottleneck. Barriers could operate on an even larger scale. The Miocene (ca. 16 Ma) alignment of the African continent is believed (Plaziat et al. 2001) to have been a major barrier to mangrove distribution between the Far East and the Atlantic, accounting for the differentiation of mangrove species between these regions.

Today's biogeographic transition between the Indochinese and Sundaic forest avifauna along the "neck" (narrowed portion) of the Thai–Malay Peninsula is distinct, with 47% of species limits examined falling in the northern part adjoining continental Thailand (Hughes et al. 2003). Among 325 native mammal species considered, the majority showed latitudinal limits along the Peninsula but not as sharply concentrated as the birds, and more than 10% of the mammals or birds occurring in both continental Thailand and Peninsular Malaysia had distribution gaps in this "neck" portion (Woodruff and Turner 2009). Such constrictions in distribution could have been due to multiple glacioeustatic highstands over the past 5 Ma that intermittently reduced land connections between continental Thailand and the

Malay Peninsula, resulting in severe habitat reduction and competition, population extinction, and faunal compression (Woodruff and Turner 2009).

2.4.3 Ecosystem Change and Vicariance

Climatic and ecosystem change events during the Late Pliocene–Early Pleistocene appear to have had an impact on the survival and evolution of many SE Asian organisms. The regression or transgression of vegetation communities, both latitudinally and vertically on continental shelves, across such episodes, as well as the emergence of new substrate or habitat, would have interrupted gene flow and exerted selective pressure on populations. This is reflected in the divergence times of organisms inferred through molecular phylogenetic studies, in an increasing list of examples (Meijaard and Groves 2006).

Pliocene–Pleistocene speciation events include, for example, divergence of the SE and East Asian bear species (Waits et al. 1999); differentiation of the slow loris *Nycticebus pygmaeus* (restricted to east of the Mekong) from the more widespread *Nycticebus bengalensis* (Zhang et al. 1993); and radiation of *Lophura* pheasants into various Sundaland species (Randi et al. 2000). *Presbytis*, the genus of Old World leaf monkeys, probably arrived in Sundaland by Late Pliocene and in response to climatic, land area, and ecological fluctuations vicariated into *P. thomasi* (Sumatra); *P. potenziani* (Mentawai Islands); *P. comata* (Java); and *P. hosei*, *P. frontata*, and *P. rubicunda* (Borneo) (Brandon-Jones 1996; Harrison et al. 2006). The mainland Asian elephant *Elephas maximus* differentiated into two haplotypes (Fernando et al. 2000). In insular SE Asia, two subspecies of orangutan, *Pongo pygmaeus*, emerged (*P. pygmaeus pygmaeus* in Borneo and *P. pygmaeus abelii* in Sumatra), with strongly divergent genetic traits probably due to their isolation on different islands (Zhang et al. 2001). These are often now accepted as distinct species (Wich et al. 2008). Divergent genetic characteristics, indicating lack of gene flow, among three forest-dwelling, Sunda-endemic, and murine rodents are reasonably explained by segregation of populations by non-forest vegetation during glacial times, and further vicariant differentiation on the various landmasses when land links were broken by rising sea levels (Gorog et al. 2004). Mitochondrial DNA variation among conspecific populations of a Bornean hill stream *Limnonectes* frog species separated by lowland also indicates that gene flow barriers exist (Zainudin in Inger and Voris (2001)). It is also likely that fluctuations in forest area corresponding to climatic oscillations during this period had impacted the ecology of *Kerivoula* woolly bats, among which active speciation was detected using a molecular clock approach for a phylogenetic analysis of mitochondrial DNA data (Khan et al. 2010).

Diversification of forest-dependent *Sundasciurus* tree squirrels within the Philippine archipelago appears to have occurred in the Pleistocene, although over much of the Malay Peninsula and Borneo the widespread *Sundasciurus lowii* and *Sundasciurus tenuis* and their localized ecological forms or close relatives seem to have diverged even earlier (den Tex et al. 2010). This apparent difference from primates and rodents, for which the Pleistocene fluctuations in SE Asia have been consequential in their speciation history, could reflect a lack of extensive forest cover over a much exposed, drier Sundaland that was much less conducive for the speciation dynamics of tree squirrels (den Tex et al. 2010).

The shrub genus *Cyrtandra* (Gesneriaceae) has not only a number of species common to Mount Kerinci (3805 m), a Pliocene–Pleistocene volcano in west Sumatra, and nearby volcanoes (representing a locally differentiated lineage) but also an entire suite of species on this mountain with nuclear ribosomal DNA traits linking them to an essentially Bornean lineage (Bramley et al. 2004). Pleistocene land connections had conceivably enabled some interchange between Sumatra and southwest/northwest Borneo. Later, a part of the Bornean lineage appears to have been isolated on the Sumatran peak by sea-level rise and upward retreat of montane vegetation in warmer, wetter times.

2.4.4 *Wallacea as a Barrier and Filter*

Wallace's Line (passing between Bali and Lombok, and also Borneo and Sulawesi) was proposed (Wallace 1863) as a boundary between the distinctive Australian and Asian faunas, and also reflects a natural boundary for many plant taxa (Whitmore 1981; van Welzen et al. 2005) (Fig. 2.11). The degree of floristic and faunal distinction between the main Sunda and Sahul regions appears to vary among taxa, dependent on which groups would have been able to disperse across different

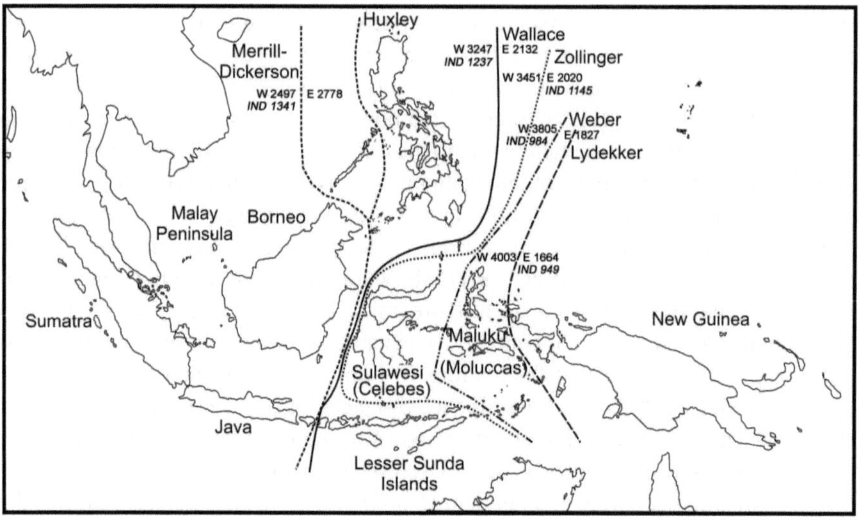

Fig. 2.11 Restriction and non-restriction of a sample of 6616 Malesian plant species to the west and east of various biogeographical boundaries in the Wallacea region: the Merrill-Dickerson, Wallace, Zollinger, Weber, and Lydekker lines. For each line, numbers of species are given that are W (restricted west of the line), E (restricted E of the line), and IND (indifferent to the line). Data from van Welzen et al. (2005)

stages of the widening Makassar Straits, which increasingly accentuated the biogeographical barrier represented by (or difficulty of dispersal across) Wallace's Line (Morley 2000). New Guinea is a "biogeographic mosaic" in that its biota has both strongly represented Asiatic or Australasian elements (Diamond 1984). Various Sunda and mainland Asiatic floristic elements that have become diverse in New Guinea probably arose from ancestors that dispersed from Halmahera and the Philippines when these were closer in the earlier Tertiary and from Sulawesi, which was much closer during the mid-Miocene (20–10 Ma) and carried Sunda plant lineages inherited from its southwest arm that had previously attached to Borneo (Morley 2003). Speciation within New Guinea was probably facilitated by increased physiographic complexity due to uplift and weathering. On the other hand, the paucity of dipterocarps east of Wallace's Line today is likely due to relatively late dipterocarp diversification in wet rainforest conditions in SE Asia only after 20 Ma (Miocene), i.e., decreased dispersal opportunities after the opening of the Makassar Straits.

In New Guinea, there was contiguity of seasonal vegetation including savanna with Australia over the Late Miocene glaciation, as sea level fell (Batchelor 1979), when initial dispersal of some organisms including marsupials is thought to have taken place (Nix and Kalma 1972). Such dry phases could have made Wallacea, the region of deeper seas and islands separating the Sunda and Sahul shelves, a more formidable barrier to cross (Mayr 1944), so marsupials were restricted to the Sahul and carnivores and primates to Sunda. This sea and climatic barrier is a biogeographical filter, restricting some groups, not all, from further dispersal. The giant freshwater prawn *Macrobrachium rosenbergii*, much cultured for the table, is naturally distributed from Pakistan to the southwest Pacific but exists as two ribosomal DNA forms separated by Huxley's Line (Huxley 1868) in the Wallacea region, passing between Bali and Lombok, through the Makassar Straits and then between Palawan and the Philippines (de Bruyn et al. 2004). This genetic divergence is reflected in hybrid vigor between forms from the same region, and poor larval survival if matings are made between different parts of the range (de Bruyn et al. 2004).

2.4.5 Isolation of Sulawesi and Other Islands

Frogs, sensitive to salt water, would only be found among landmasses separated by seas if there were historical periods of land contiguity permitting dispersals. There is a greater proportion of shared species among the Sundaland territories (Malay Peninsula, Sumatra and Borneo) than any of these with Sulawesi, which was never truly on the continental shelf (except when its southwestern arm was joined to southeast Borneo before the mid-Eocene) (Inger and Voris 2001). Snakes, too, are less similar between Sulawesi and any of the Sunda landmasses than among the Sunda territories themselves (Inger and Voris 2001).

Besides having less in common with continental islands (those that have continental shelf or past land connections with the main continent), long isolated islands such as Sulawesi have other interesting features (de Vos et al. 2007). Their organismal subsets are "disharmonic" (unbalanced in composition), missing out some groups typical in the nearest continental source areas (e.g., having few or no carnivores, with mostly herbivores such as proboscideans, hippopotami, and cervids). High levels of endemism and frequent dwarfism or gigantism exist. Endemicity in Sulawesi includes 98% of its mammals (excluding bats), 76% of its amphibians, nearly 30% of its swallowtail butterflies (Whitten et al. 1987), and 95% of cicadas (Duffels 1990). In Sulawesi, a pygmy *Stegodon* (*S. sompoensis*), a pygmy elephant (*Elephas celebensis*), and a giant tortoise *Geochelone atlas* were present in the Late Pliocene to Pleistocene (van den Bergh 1999). Flores, south of Sulawesi, had a giant rat *Hooijeromys nusatenggara*, the pygmy *Stegodon sondaari*, and a *Colossochelys* giant tortoise (de Vos et al. 2007). The Wallacea islands of Timor and Sumba, and the Philippines, had other extinct dwarf forms too (de Vos and Bautista 2003; Croft et al. 2006).

The evolution of very specialized or unusual taxa appears to be due to extreme isolation, such as the anao, *Bubalus depressicornis*, and babirusa, *Babyrousa babyrussa*, both well-known Sulawesi curiosities (Whitten et al. 1987). It is possible that plants, too, show this influence. For example, *Porterandia* (Rubiaceae), an essentially Sundaland genus of trees, has a single outlier species endemic to Sulawesi (*P. celebica*) with conspicuous boat-shaped to hood-like inflorescence bracts (fused along one or both margins) (Fig. 2.12) not found in the other 21 species, which have non-fused bracts (Zahid 2004; Zahid and Wong 2010).

Farther north, Palawan's impoverished fauna (Esselstyn et al. 2004: 58 extant native mammals, compared to 285 on Borneo) is similarly interesting. Located at the NE transition between the Sunda and Wallacea biogeographic regions, the relative isolation of Palawan from Borneo over the LGM appears to have resulted in its high mammalian endemism and an absence of common closed forest species of the Sunda shelf that could only have developed following the LGM (Piper et al. 2010). Notwithstanding, even the tiger had reached Palawan, its eventual extinction probably due to a combination of habitat loss, reduced prey populations, and hunting pressure (Piper et al. 2008).

2.4.6 Dispersal Events and Speciation

The importance of biogeographical barriers and vicariance in restricting gene flow among populations is suggested by congruence between phylogenetic or phylogeographical patterns of organisms and populations and reconstructed geological and area relationships. Incongruence may indicate that other processes, chiefly dispersal, could have influenced the perceived distribution patterns. Dispersals are "sweepstake" when no special ability to disperse over a barrier is apparent but some species get across by chance or "filter" when some organisms easily cross a barrier (perhaps aided by adaptations such as wind-dispersible light seeds or ability to swim) but others cannot (perhaps restricted by heavy seeds and intolerance of seawater conditions, and so on).

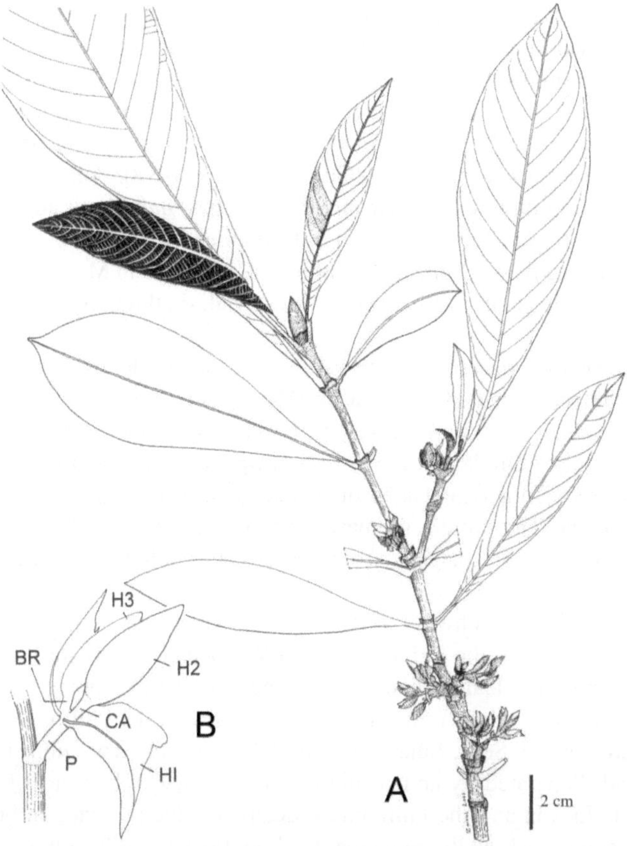

Fig. 2.12 *Porterandia celebica*, a Sulawesi oddity. (**a**) Leafy branch with bisexual inflorescences. (**b**) Diagram of an inflorescence from (**a**) showing the peduncle (P), central inflorescence axis above this (CA), and only one developed primary branch (BR), with the corresponding hood-like structures formed by fused bracts marked as H1 (torn), H2 (subtends the undeveloped parts on the central axis), and H3 (subtends the undeveloped parts on the primary branch). Odd features among plants sometimes require detailed examination to appreciate, hence they may not be as conspicuous as those in many animals. Drawing by Zainal Mustafa, reproduced with permission of the Forest Research Centre, Sandakan, Malaysia

Australian elements have dispersed across Wallace's Line into Sundaland mostly during low sea levels in the Miocene–Pliocene period (Hall 1996, 1998), after formation of the Lesser Sundas and the Banda Arc and uplift of New Guinea. These have included easily dispersed lowland species (such as those with light-winged seed and ability to survive harsh conditions), montane taxa able to disperse along the cool corridors of mountain ranges, or chance dispersals. Common in Australia before the Miocene, *Casuarina*-type pollen only appeared in SE Asia after the Australian collision with Asia during this epoch. There was increased pollen abundance of Myrtaceae (an ancient Australian family) in the Malay Basin, Natuna Sea,

Sarawak, and Mahakam Delta only from about 17 Ma (Morley 2000). Some others became more widespread and then became extinct in places, such as *Stenochlaena milnei*-type fern spores that first appeared in Java, Sarawak, and the Mahakam Delta at 9.5 Ma (earliest Late Miocene), then spread to south China, Myanmar, and India, but became extinct in Sundaland in the earliest Pleistocene; the species is now restricted to the Philippines and New Guinea (Morley 1998). Pollen of Australasian montane plants resembling *Dacrycarpus* (*Podocarpus*) *imbricatus* and *Phyllocladus* appeared 3.5–1.6 Ma (Early Pliocene) in Borneo and Java, and in Borneo and the Philippines, respectively. *Podocarpus* pollen had appeared in Myanmar by the early Pleistocene and the genus is still found in west Malesia through Indochina (Morley 2000).

A phylogenetic analysis of the SE Asian stream-loving dragonfly genus *Macromia* indicated that Halmahera and New Guinea taxa were derived from Sulawesi, which in turn were closest to Sundaland forms (van Tol 1994). This implies dispersal of Sundaland forms to Sulawesi prior to separation of its south-western arm from Borneo, and subsequent dispersal (island hopping) to Halmahera and New Guinea. The pattern of marine depths do not rule out the intermittent existence of interspersed islands from Borneo, through Java, to Sulawesi (Hall 2009).

Another dispersal pathway must exist between the Asian mainland and Australasia via Taiwan, the Philippines, and Wallacean islands farther south. This is part of a series of volcanic islands associated with the Philippine tectonic plate's western margin, stretching from Japan through Taiwan, the Philippines, islands in the Sulu and Celebes Seas, Sulawesi, and Halmahera to New Guinea and toward New Zealand. Separated by an oceanic trench, the lack of historical land connection between Taiwan and the Philippines means that the presence in both areas of the same species, such as the montane bamboo *Yushania niitakayamensis* (Jacobs 1972; Wong 2004), must have occurred via dispersal. Likewise, the distribution of *Dracaena multiflora* (Asparagaceae) and *Selaginella tamariscina* (Selaginellaceae) in both the Philippines and on vestiges of volcanic rims off east Sabah (Wong et al. 1999) and of *Gardenia mutabilis* (Rubiaceae) in the Philippines and Sulawesi (Wong and Low 2011) is attributable to dispersal along this track. Molecular relationships indicate that the alpine plant genus *Oreomyrrhis* (Apiaceae) migrated from Australia northward along this general route (Chung et al. 2005) although other taxa could have been dispersed out of mainland Asia in the reverse direction (Wu et al. 2009).

2.5 Interface with Ecological Processes: Physiographic–Edaphic Diversification

Land area has a general relationship with species richness. For example, among islands on the Sunda shelf, a logarithm plot of mammal species tallies and areas yields a positive linear relationship (Heaney 1984). However, landform evolution and the formation of different rock types and substrates are also important, as they contribute both landscape level and local physiographic and edaphic complexity. Eco-geographic partitioning of genetic variation can cause population and taxonomic divergences. Because larger islands typically have more landform variation than small ones, it is not surprising that the major islands of Malesia have endemism levels correlated to their size (Roos et al. 2004; Raes et al. 2009). The distribution of *Calamaria*, a fossorial-secretive group of snakes that could not have "island hopped" on floating trees across seas, on different landmasses no doubt originated during past episodes of land contiguity, but the many island endemics (11 of 15 species in Sumatra, 14 of 22 in Borneo, 9 of 10 in Java, 7 of 8 in Sulawesi) and the occurrence of montane endemics in Sumatra and Borneo probably arose through local diversification in response to terrain variation and ecosystem change from the Miocene to the Quaternary (Inger and Voris 2001).

2.5.1 Landform

In the highlands, organisms have adapted to variation with elevation in atmospheric pressures, radiation, humidity, temperature, and associated changes. They form natural communities including the range of forests from lowland to montane and alpine types. The highest mountains in SE Asia exist in geologically young and physiographically complex formations, in north Borneo and New Guinea, where biological richness has reached very high levels.

Mount Kinabalu (4094 m), a Pliocene–Pleistocene granitic intrusion at the northeastern tip of the Crocker range, is the highest peak in Borneo. Its elevation and edaphic range sustain vegetation formations from warm lowland forests to alpine scrub and communities on ultramafic outcrops. A precipitous topography (Fig. 2.13) with frequent landslides, and past disturbance history from Pleistocene climate change and drought episodes, apparently exerted high selective pressure on organisms and partitioned species continuities, fueling adaptive radiation. Some 5000 species of vascular plant species have been recorded, around 14% of the estimated 42,000 species in the entire Malesian region (Beaman and Anderson 2004). Among *Elatostema* herb species (Urticaceae), high ultramafic outcrops have high endemicity, and many phylogenetically more derived taxa occur at mid-elevation where past glaciation has crafted deep valleys isolated by high sharp ridges (Beaman 2001). It is also likely, as on the Sumatran Mount Kerinci, of comparable age, that relict parts of lineages from elsewhere have persisted and radiated (Bramley et al. 2004). Along Kinabalu's elevation gradient up to 2300 m, 283 species of 55 genera of ground litter ants have been documented (Bruhl et al. 1999), of which some 74% were restricted to one site (i.e., there was high species turnover along the slope). Some plant species in the subalpine community are endemics closely related to (apparently vicariant with) those in lower zones, e.g., *Leptospermum recurvum* (Myrtaceae), *Dacrycarpus kinabaluensis*, and *Dacrydium gibbsiae* (Podocarpaceae) are related, respectively, to *L. javanicum*, *Dacrycarpus imbricatus*, and *Dacrydium elatum* found lower down. Yet vicariance by itself may not entirely explain this. Vertically differentiated parapatric species could have resulted from competitive displacement of less successful

Fig. 2.13 Sharp ridges and deeply incised valleys at mid-elevation, Crocker Range, Sabah, Borneo. Photo courtesy of M. Sugumaran

alleles and species which arise but have highly similar resource-use profiles as (and so directly compete with) existing taxa (Kitayama 1996).

The high mountain ridges of New Guinea represent a youthful geomorphology, with sharp crests separated by deeply incised valleys (Löffler 1982). Peaks along the central ranges rise over 4000 m with the highest, Mount Jaya, reaching 5000 m; a number reaches the snow line. Limestone karst landscapes and volcanoes add to the range of landforms and substrate diversity. New Guinea has over 13,000 vascular plant species compared to Borneo's estimate of over 14,000, but New Guinea endemism levels are much higher, averaging 46% compared to Borneo's 28%, based on some 15% of the recently revised Malesian flora (Roos et al. 2004). This probably resulted from very high rates of radiation on New Guinea, where, for example, the 1000–3000 m zone has most of the 400 species of Ericaceae (of 777 species in Malesia), as well as most species of birds-of-paradise and bowerbirds (Heads 2003).

2.5.2 Specialized Substrates, Ecological Islands

Throughout SE Asia, tectonic, landform, and ecological processes have produced a great variety of rocks and substrates, ranging from granite, calc-alkaline rocks (e.g., andesite, dacite, and basalt), ultramafics derived from uplifted seafloor material (with high levels of nickel, chromium, iron, lead, zinc, manganese, pyrhotite, etc.), metamorphics, quartz reefs (Fig. 2.14), sand terraces, carbonate karst, gigantic peat domes, and alluvial plains. Many such areas are sharply differentiated from

Fig. 2.14 The Klang Gates quartz reef northeast of Kuala Lumpur in the Malay Peninsula. Several flowering plant species are endemic to this ecological island. Photo courtesy of D.K.P. Lee

surrounding substrates and form ecological (habitat) islands, the way that mountain peaks and tower karst formations are islands (Wong et al. 2010). Factors that accentuate such island environments include, for example, reduced soil development or nutrient limitation (Brearley 2005) or texture-related characteristics such as drought or waterlogging, and differential selection of traits that promote the generation of endemic forms (Brady et al. 2005).

These substrates can support distinctive vegetation types and communities with characteristic species and endemics. For example, "endemic ultramafic" trees in northern Borneo include the lowland gigantic *Borneodendron aenigmaticum* (Euphorbiaceae) (Fig. 2.15) and the smaller *Scaevola micrantha* (lowlands), *Scaevola chanii* (montane, Kinabalu) (Goodeniaceae), *Pittosporum linearifolium* (Kinabalu), and *Pittosporum silamense* (Silam) (Pittosporaceae). Specialist Bornean *kerangas* plants include the dipterocarp trees *Cotylelobium burckii*, *Dryobalanops fusca*, *Shorea retusa*, *Shorea revoluta*, *Vatica compressa* (Dipterocarpaceae), the she-oak *Quercus kerangasensis* (Fagaceae), and the rattan *Calamus ashtonii* (Arecaceae). Coastal *kerangas* and peat swamps often occur together and their communities are bounded by the sea and more landward mixed dipterocarp forest on better soils, which probably explains the existence of endemics shared by *kerangas* and peat swamp forests (Wong 1998), e.g., the trees *Dryobalanops rappa*, *Shorea inaequilateralis*, *Shorea pachyphylla* (Dipterocarpaceae), *Dactylocadus stenostachys* (Melastomataceae), and the large liana *Aidia dilleniacea* (Rubiaceae).

Fig. 2.15 Restricted to ultramafic sites in north Borneo, *Borneodendron aenigmaticum* is lone representative of its genus, has whorled leaves, and yields a red sap unusual in the plant family Euphorbiaceae. Photo courtesy of M. Sugumaran

2.6 Change in Past, Present, and Future

While taxonomic inventories enumerate species richness, biogeographical comparisons and analyses help identify areas of special richness, their characteristics, and their extents. In particular, the ability to express which biogeographic areas harbor least similarity in their biotas allows us to maximize complementarity among potential conservation areas harboring special diversity (Platnick 1992). The development of indices to estimate overall phylogenetic diversity, not just relying on species counts, permits estimation of taxonomic distinctness and contributes to our ability to rank areas by priority (Faith 1994). Various approaches for identifying biologically and evolutionarily important regions and areas according to biodiversity richness, endemicity, or phylogenetic distinctness fall in the realm of biogeography (Crisci et al. 2003).

These and population characteristics of particular species, including some knowledge of genetic processes, how these vary, and what factors influence them, inform conservation choices (Woodruff 1990). Phylogeographic studies that reveal the evolving genetic traits of species, such as with the giant freshwater prawn and even mangrove trees that have distinct haplotypes in different parts of SE Asia (above), can save us resources and circumvent problems in managing these species. Although carefully mixing genetic traits could yield the products we plan, conservation of genetic diversity maintains long-term adaptability of populations in the face of environmental changes (Willi et al. 2006) and is desirable in natural landscapes. In managing dwindling orangutan populations, human-mediated breeding programs could have mixed Sumatran and Bornean haplotypes in Central Kalimantan (Kanthaswamy and Smith 2002), so breeding programs that could alter such diversity should be carefully planned. There is also strong genetic differentiation between Bornean and Sumatran clouded leopards, possibly even indicating specific distinction, and both are threatened (Wilting et al. 2007). Conservation programs for endangered species will benefit from phylogeographic studies that help identify populations or individuals for potential outbreeding. However, long-captive populations may have developed genetic traits that could hinder adaptation of offspring to wild conditions.

The species-rich ecological assemblages of SE Asia have been molded via both historical and ecological processes (Morley 2000; Wright 2002, 2007). Over both short and long timeframes, environmental inconstancy and heterogeneity play a fundamental part in shaping speciation and richness. Climate change has been important in past transformations of SE Asian ecosystems, which can bounce back, transform, or go extinct. Currently, however, the evergreen rainforest communities of Sundaland are likely to be much reduced in extent compared to their past distribution over the LGM, so this refugial state is highly vulnerable to disturbance (Canon et al. 2009; Woodruff 2010). The processes of change in the present time, however, are largely human mediated and unprecedented in scale and impact. A modern appreciation of biogeography now cannot avoid touching on the apparent world domination over nature by one modern species, biological introductions and invasions that see organisms intrude into and modify communities far from their origin, disease pandemics carried over the globe by our incessant need to mingle, and the decimation of natural systems that could only have developed over geological eras through a rich interplay of natural processes.

SE Asia's lowland rainforests were once vast expanses of an incredibly rich ecological formation. These forests are like genetic libraries available to us for carefully extracting useful information and products, and a better understanding of these ecosystems will improve our ability to carefully manage them in sustainable ways. New technologies in genetics present an opportunity to develop novel ways of assessing patterns in biodiversity and may be used to improve conservation management. We have lost much in SE Asia: the worst in terms of species, genetic resources, and long-term adaptability of natural systems (not to mention "opportunities") due to loss of habitat, when natural land cover is degraded or transformed, as in the seemingly inexorable sweep of plantation agriculture over landscapes, often following from a lack of coordinated land-use management. Biogeographic studies have helped and will continue to improve our understanding of the interrelationships between environments and ecosystems, past and present, and appreciate the agencies of change. The coming together of the sciences (Lieberman 2005) has enriched our biogeographical perspective and has provided new insights into the observation of nature, how it has changed through time and how we can minimize the losses brought about by anthropogenic activities. What this cannot do is to help us reassemble such an extensive and diverse array of organisms, genetic traits, and natural compounds in as short a time as we have taken to extinguish them.

Acknowledgments Parts of this review were enabled by a Mercer Fellowship from Harvard University's Arnold Arboretum, July to November 2007. I am grateful to Bob Cook, Stuart Davies, Chuck Davis, Dave Boufford, Kanchi Gandhi, Judith Warnement, Gretchen Wade, William Tootle, Lisa DeCesare, Emily Wood, Zhang Wenheng (Harvard University); Richard Primack (Boston University); John Beaman (then Royal Botanic Gardens, Kew); Chris Yusheng Liu (East Tennessee State University); Haji Mohamed Abdul Majid, Amru Nasrulhaq Boyce, A.L. Lim (University of Malaya); Y.F. Lee, R.C. Ong, Anuar Mohammad (Forest Research Centre, Sandakan); C.L. Chan (Natural History Publications, Borneo); S.C. Chin and Christina Soh (Singapore Botanic Gardens) for facilitating my studies and help with literature. I thank the Earl of Cranbrook and James Richardson (Royal Botanic Garden Edinburgh) for comments on a manuscript version of this chapter.

References

Aleva GJJ, Bon EH, Nossin JJ, Sluiter WJ (1973) A contribution to the geology of part of the Indonesian tinbelt: the sea areas between Singkep and Bangka Islands and around the Karimata Islands. Geol Soc Malays Bull 6:257–271

Alfaro ME, Karns DR, Voris HK, Abernathy E, Sellins SL (2004) Phylogeny of *Cerberus* (Serpentes: Homalopsinae) and phylogeography of *Cerberus rynchops*: diversification of a coastal marine snake in Southeast Asia. J Biogeogr 31:1277–1292

Alvarez LW, Alvarez W, Asaro F, Michel HV (1980) Extraterrestrial cause for Cretaceous-Tertiary extinction. Science 208:1095–1108

Ashton PS (1972) The Quaternary geomorphological history of western Malesia and lowland forest phytogeography. In: Ashton P, Ashton M (eds) The quaternary Era in Malesia, Misc Ser No. 13. Department of Geography, University of Hull, Aberdeen. pp 35–49

Ashton PS (1982) Dipterocarpaceae. In: van Steenis CGGJ (ed) Flora Malesiana Series I, vol 9. Martinus Nijhoff, The Hague. pp 237–552

Ashton PS, Gunatilleke CVS (1987) New light on the plant geography of Ceylon 1. Historical plant geography. J Biogeogr 14:249–285

Azuma N, Ogata K, Kikuchi T, Higashi S (2006) Phylogeography of Asian weaver ants, *Oecophylla smaragdina*. Ecol Res 21:126–136

Barber AJ, Crow MJ, Milsom JS (2005) Sumatra: geology, resources and tectonic evolution. Geol Soc Lond Mem 31. 290 p

Batchelor BC (1979) Discontinuously rising Late Cainozoic eustatic sea-levels, with special reference to Sundaland, Southeast Asia. Geol Mijnbouw 58:1–20

Beaman RS (2001) Phylogeny and biogeography of *Elatostemma* (Urticaceae) from Mount Kinabalu. Sabah Parks Nat J 4:71–93

Beaman JH, Anderson C (2004) The plants of Mount Kinabalu. 5. Dicotyledon Families Magnoliaceae to Winteraceae. Natural History Publications (Borneo), Kota Kinabalu

Bird MI, Taylor D, Hunt C (2005) Palaeoenvironments of insular Southeast Asia during the Last Glacial Period: a savanna corridor in Sundaland? Quatern Sci Rev 24:2228–2242

Bird MI, Pang WC, Lambeck K (2006) The age and origin of the Straits of Singapore. Palaeogeogr Palaeoclim Palaeoecol 241:531–538

Brady KU, Kruckeberg AR, Bradshaw HD (2005) Evolutionary ecology of plant adaptation to serpentine soils. Annu Rev Ecol Evol Syst 36:243–266

Bramley GLC, Pennington RT, Zakaria R, Tjitrosoedirdjo SS, Cronk QCB (2004) Assembly of tropical plant diversity on a local scale: *Cyrtandra* (Gesneriaceae) on Mount Kerinci, Sumatra. Biol J Linnean Soc 81:49–62

Brandon-Jones D (1996) The Asian Colobinae (Mammalia: Cercopithecidae) as indicators of quaternary climatic change. Biol J Linnean Soc 59:327–350

Brandon-Jones D (2001) Borneo as a biogeographic barrier to Asian-Australasian migration. In: Metcalfe I, Smith JMB, Morwood M, Davidson I (eds) Faunal and floral migrations and evolution in SE Asia-Australasia. Balkema, Lisse. pp 365–372

Brearley FQ (2005) Nutrient limitation in a Malaysian ultramafic soil. J Trop Forest Sci 17(4): 596–609

Bruhl CA, Mohamed M, Linsenmair KE (1999) Altitudinal distribution of leaf litter ants along a transect in primary forests on Mount Kinabalu, Sabah, Malaysia. J Trop Ecol 15:265–277

Bunopas S, Wasson JT, Vella P, Fontaine H, Hada S, Burrett C, Suphajunya T, Khositanont S (1999) Catastrophic losses, mass mortality and forest fires suggest that a Pleistocene cometary impact in Thailand caused the Australasian tektite field. J Geol Soc Thai 1:1–17

Burrett C, Duhig N, Berry R, Varne R (1991) Asian and south-western Pacific continental terranes derived from Gondwana, and their bogeographic significance. Aust Syst Bot 4:13–24

Cannon CH, Manos PS (2003) Phylogeography of the Southeast Asian stone oaks (*Lithocarpus*). J Biogeogr 30:211–226

Cannon CH, Morley RJ, Bush ABG (2009) The current refugial rainforests of Sundaland are unrepresentative of their biogeographic past and highly vulnerable to disturbance. Proc Natl Acad Sci USA 106:11188–11193

Chatterjee S, Rudra DK (1996) KT events in India: impact, rifting, volcanism, and dinosaur extinction. Mem Qld Mus 39:489–532

Chung KF, Peng CI, Downie SR, Spalik K, Schaal BA (2005) Molecular systematics of the trans-pacific alpine genus *Oreomyrrhis* (Apiaceae): phylogenetic affinities and biogeographic implications. Am J Bot 92:2054–2071

Corner EJH (1960) The Malayan flora. In: Proceedings of the centenary and bicentenary congress of biology, University of Singapore, Singapore

Cranbrook Earl of (2000) Northern Borneo environments of the past 40,000 years. Sarawak Mus J 55:61–109

Cranbrook Earl of (2010) Late quaternary turnover of mammals in Borneo: the zooarchaeological record. Biodivers Conserv 19:373–391

Cranbrook Earl of, Piper PJ (2008) Sarawak through the Ice Ages to present time: environmental change and human impacts on the past and present distribution of mammals. In: Proceedings of the regional conference on biodiversity & conservation in tropical planted forests in Southeast Asia, Natural History Publications (Borneo), Kota Kinabalu

Cranbrook Earl of, Piper PJ (2009) Borneo records of Malay tapir, *Tapirus indicus* Desmarest: a zooarchaeological and historical review. Int J Osteoarchaeol 19:491–507. doi:10.1002/oa.1015

Crisci JV, Katinas L, Posadas P (2003) Historical biogeography. An introduction. Harvard University Press, Cambridge, MA

Croft DA, Heaney LR, Flynn JJ, Bautista AP (2006) Fossil remains of a new, diminutive Bubalus (Artiodactyla: Bovidae: Bovini) from Cebu Island, Philippines. J Mammol 87:1037–1051

de Bruyn M, Wilson JA, Mather PB (2004) Huxley's Line demarcates extensive genetic divergence between eastern and western forms of the giant freshwater prawn, *Macrobrachium rosenbergii*. Mol Phylogenet Evol 30:251–257

den Tex R-J, Thorington R, Maldonado JE, Leonard JA (2010) Speciation dynamics in the SE Asian tropics: putting a time perspective on the phylogeny and biogeography of Sundaland tree squirrels, *Sundasciurus*. Mol Phylogenet Evol 55:711–720

de Vos J, Bautista A (2003) Preliminary notes on the vertebrate fossils from the Philippines. *Semantics and Systematics:* Philippine Archaeology, Proceedings of the Society of Philippine Archaeologists, vol 1, pp 42–62

de Vos J, van den Hoek Ostende LW, van den Bergh GD (2007) Patterns in insular evolution of mammals: a key to island palaeogeography. In: Renema W (ed) Biogeography, time and place: distributions, barriers, and islands. Springer, New York, NY. pp 315–346

Diamond J (1984) Biogeographic mosaics in the Pacific. In: Radovsky FJ, Raven PH, Sohmer SH (eds) Biogeography of the tropical pacific. Bishop Museum Special Publications No. 72. Bishop Museum, Honolulu. pp 1–74

Duffels JP (1990) Biogeography of Sulawesi cicadas (Homoptera: Cicadoidea). In: Knight WD, Holloway JD (eds) Insects and the rain forests of South East Asia (Wallacea). Royal Entomological Society of London, London. pp 63–72

Emery KO, Uchupi E, Sunderland J, Uktolseja HL, Young EM (1972) Geological structure and some water characteristics of the Java Sea and adjacent continental shelf. CCOP Tech Bull 6:197–223

Enfield DB (1992) Historical and prehistorical overview of El Niño/Southern Oscillation. In: Diaz HF, Markgraf V (eds) El Niño: historical and paleoclimatic aspects of the southern oscillation. Cambridge University Press, Cambridge. pp 96–117

Esselstyn JA, Widmann P, Heaney LR (2004) The mammals of Palawan Island, Philippines. In: Proc. Biol. Soc. Washington 117:271–302

Faith DP (1994) Phylogenetic pattern and the quantification of organismal diversity. Philos Trans R Soc Lond B 345:45–58

Ferguson DK (1993) The impact of Late Cenozoic environmental changes in East Asia on the distribution of terrestrial plants and animals. In: Jablonski NG, Chak-Lam S (eds) Evolving landscapes and evolving Biotas of East Asia since the mid-tertiary. Proceedings of the 3rd conference on the evolution of the East Asian environment. University of Hong Kong, Hong Kong. pp 145–196

Fernando P, Pfrender ME, Encalada S, Lande R (2000) Mitochondrial DNA variation, phylogeography and population structure of the Asian elephant. Heredity 84:362–372

Flenley JR (1979a) The equatorial rain forest: a geological history. Butterworths, London

Flenley JR (1979b) The late quaternary vegetational history of the equatorial mountains. Prog Phys Geogr 3:487–509

Flenley JR (1984) Late quaternary changes of vegetation and climate in the Malesian mountains. Erdwissenschaft Forsch 18:261–267

Friis EM, Pederson KR, Crane PR (1994) Angiosperm floral structures from the Early Cretaceous of Portugal. In: Endress PK, Friis EM (eds) Early evolution of flowers, plant systematics and evolution, Suppl 8. Springer, Vienna. pp 31–49

Gathorne-Hardy FJ, Syaukani, Davies RG, Eggleton P, Jones DT (2002) Quaternary rainforest refugia in south-east Asia: using termites (Isoptera) as indicators. Biol J Linnean Soc 75: 453–466

Goh KC (2005) The climate of Southeast Asia. In: Gupta A (ed) The physical geography of Southeast Asia. Oxford University Press, Oxford. pp 80–93

Gorog AJ, Sinaga MH, Engstrom MD (2004) Vicariance or dispersal? Historical biogeography of three Sunda shelf murine rodents (*Maxomys surifer*, *Leopoldamys sabanus* and *Maxomys whiteheadi*). Biol J Linnean Soc 81:91–109

Hall R (1996) Reconstructing Cenozoic South East Asia. In: Hall R, Blundell DJ (eds) Tectonic evolution of Southeast Asia. Geological Society Special Publication 106. pp 153–184

Hall R (1998) The plate tectonics of Cenozoic South East Asia and the distribution of land and sea. In: Hall R, Holloway J (eds) Biogeography and geological evolution in South East Asia. Bakhuys Publishers, Amsterdam. pp 99–131

Hall R (2009) Southeast Asia's changing palaeogeography. Blumea 54:148–161

Harley MM, Morley RJ (1995) Ultrastructural studies of some fossil and extant palm pollen, and the reconstruction of the biogeographical history of the subtribes Iguanurinae and Calaminae. Rev Palaeobot Palynol 85:153–82

Harrison T, Krigbaum J, Manser J (2006) Primate biogeography and ecology on the Sunda Shelf Islands: a paleontological and zooarchaeological perspective. In: Lehman SM, Fleagle JG (eds) Primate biogeography. Springer, New York, NY. pp 331–374

Heads M (2003) Ericaceae in Malesia: vicariance biogeography, terrane tectonics and ecology. Telopea 10:311–449

Heaney LR (1984) Mammalian species richness on islands on the Sunda Shelf, Southeast Asia. Oecologia 61:11–17

Heaney LR (1991) A synopsis of climatic and vegetational change in Southeast Asia. Climate Change 19:53–61

Heil A, Goldammer JG (2001) Smoke-haze pollution; a review of the 1997 episode in Southeast Asia. Reg Environ Change 2:24–37

Hildebrand AR, Penfield GT, King DA, Pilkington M, Camargo ZA, Jacobsen SB, Boynton WV (1991) Chicxulub Crater: a possible Cretaceous/Tertiary boundary impact crater on the Yucatan Peninsula, Mexico. Geology 19:867–871

Hope GS (2001) Environmental change in the Late Pleistocene and later Holocene at Wanda site, Soroako, South Sulawesi, Indonesia. Palaeogeogr Palaeoclim Palaeoecol 171:129–145

Hughes JB, Round PD, Woodruff DS (2003) The Sundaland–Asian faunal transition at the Isthmus of Kra: an analysis of resident forest bird species distributions. J Biogeogr 30:569–580

Huxley TH (1868) On the classification and distribution of the Alectoromorphae and Heteromorphae. Proc Zool Soc Lond 1868:294–319

Inger RF, Voris HK (2001) The biogeographical relations of the frogs and snakes of Sundaland. J Biogeogr 28:863–891

Jablonski NG, Whitfort MJ (1999) Environmental change during the Quaternary in East Asia and its consequences for mammals. Records West Austr Mus 57:307–315

Jackson ST (2006) Vegetation, environment, and time: the origination and termination of ecosystems. J Vegetat Sci 17:549–557

Jacobs M (1972) The plant world on Luzon's highest mountains. Rijksherbarium, Leiden

Jacobs M (1974) Botanical panorama of the Malesian archipelago (vascular plants). In: Natural resources in humid tropical Asia. UNESCO, Paris. pp 263–294

Johnson KR, Ellis B (2002) A tropical rainforest in Colorado 1.4 million years after the Cretaceous-Tertiary boundary. Science 296:2379–2383

Kanthaswamy S, Smith DG (2002) Population subdivision and gene flow among wild orangutans. Primates 43:315–327

Karns DR, O'Bannon A, Voris HK, Weigt LA (2000) Biogeographical implications of mitochondrial DNA variation in the Bockadam snake (*Cerberus rynchops*, Serpentes: Homalopsinae) in Southeast Asia. J Biogeogr 27:391–402

Khan FAA, Solaris S, Swier VJ, Larsen PA, Abdullah MT, Baker RJ (2010) Systematics of Malaysian woolly bats (Vespertilionidae: *Kerivoula*) inferred from mitochondrial, nuclear, karyotypic, and morphological data. J Mammal 91:1058–1072

Kirby GAK, Morley RJ, Humphreys B, Matchette-Downes CJ, Sarginson MJ, Lott GK, Nicholson RA, Yulihanto B, Widiastuti R, Karmajaya S, Fitris F, Sofyan S, Wijaya S (1993) A revaluation of the regional geology and hydrocarbon prospectivity of the onshore Central North Sumatra Basin. In: Proceedings of the 22nd convention, Jakarta, Indonesian Petroleum Association

Kitayama K (1996) Patterns of species diversity on an oceanic versus a continental island mountain: a hypothesis on species diversification. J Vegetat Sci 7:879–888

Kuhnt W, Holbourn A, Hall R, Zuvela M, Käse R (2004) Neogene history of the Indonesian throughflow. AGU Geophys Monogr 149:299–320

Lambeck K, Chappell J (2001) Sea level change through the last glacial cycle. Science 292: 679–686

Lee M-Y, Wei K-Y (2000) Australasian microtektites in the South China Sea and the West Philippine Sea: implications for age, size and location of the impact crater. Meteorit Planet Sci 35:1151–1155

Leong TM, Grismer LL, Mumpuni (2003) Preliminary checklists of the herpetofauna of the Anambas and Natuna islands (South China Sea). Hamadryad 27:165–174

Li X, Sun X (1999) Palynological records since the Last Glacial Maximum from a deep sea core in the southern South China Sea. Quatern Sci 6:526–535

Liao P-C, Havanond S, Huang S (2007) Phylogeography of *Ceriops tagal* (Rhizophoraceae) in Southeast Asia: the land barrier of the Malay Peninsula has caused population differentiation between the Indian Ocean and South China Sea. Conserv Genet 8:89–98

Lieberman BS (2005) Geobiology and paleobiogeography: tracking the coevolution of the Earth and its biota. Palaeogeogr Palaeoclim Palaeoecol 219:23–33

Liu J, Tang L, Qiao Y, Head MJ, Walker D (1986) Late Quaternary vegetational history at Menghai, Yunnan Province, Southwest China. J Biogeogr 13:399–418

Löffler E (1982) Landforms and landform development. In: Gressitt JL (ed) Biogeography and ecology of New Guinea. Dr W Junk Publishers, The Hague

Lourie SA, Green DM, Vincent ACJ (2005) Dispersal, habitat differences, and comparative phylogeography of Southeast Asian seahorses (Syngnathidae: *Hippocampus*). Mol Ecol 14:1073–1094

Maloney BK, McCormac FG (1995) A 30,000 year pollen and radiocarbon record from highland Sumatra as evidence for climate change. Radiocarbon 37:181–190

Mayr E (1944) Wallace's line in the light of recent zoogeographic studies. Q Rev Biol 19:1–14

Medway Lord (1960) The Malay tapir in Quaternary Borneo. Sarawak Mus J 9:364–367

Medway Lord (1972) The Quaternary mammals of Malesia: a review. In: Ashton P, Ashton M (eds) The Quaternary Era in Malesia. Transactions of the 2nd Aberdeen-Hull symposium on Malesian ecology, series 13. University of Hull Department of Geography Misc, Aberdeen. pp 63–98

Meijaard E, Groves CP (2006) The geography of mammals and rivers in mainland Southeast Asia. In: Lehman SM, Fleagle JG (eds) Primate biogeography. Springer, New York, NY. pp 305–329

Metcalfe I (1998) Palaeozoic and Mesozoic geological evolution of the SE Asian region: multidisciplinary constraints and implications for biogeography. In: Hall R, Holloway JD (eds) Biogeography and geological evolution of SE Asia. Backhuys Publishers, Leiden. pp 25–41

Metcalfe I (2002) Permian tectonic framework and paleogeography of SE Asia. J Asian Earth Sci 20:551–566

Morley R (1991) Tertiary stratigraphic palynology in South-East Asia: current status and new directions. Bull. Geol. Soc. Malaysia 28:1–36

Morley R (1998) Palynological evidence for Tertiary plant dispersals in the South East Asia region in relation to plate tectonics and climate. In: Hall R, Holloway J (eds) Biogeography and geological evolution in South East Asia. Bakhuys Publishers, Amsterdam. pp 211–234

Morley RJ (2000) Origin and evolution of tropical rain forests. Wiley, Chichester

Morley RJ (2003) Interplate dispersal paths for megathermal angiosperms. Perspect Plant Ecol Evol Syst 6(12):5–20

Morley R, Flenley JR (1987) Late Cainozoic vegetational and environmental changes in the Malay Archipelago. In: Whitmore TC (ed) Biogeographical evolution of the Malay Archipelago, Oxford monographs in biogeography no. 4. Oxford Scientific Publications, Oxford. pp 50–59

Muller J (1966) Montane pollen from the Tertiary of N. W. Borneo. Blumea 14:231–235

Muller J (1972) Palynological evidence for change in geomorphology, climate and vegetation in the Mid-Pliocene of Malesia. In: Ashton P, Ashton M (eds) The Quaternary era in Malesia. Transactions of the 2nd Aberdeen-Hull symposium on Malesian ecology, series no. 13. University of Hull Department of Geography Misc, Aberdeen. pp 6–34

Muller J (1981) Fossil pollen records of extant angiosperms. Bot Rev 47:1–142

Nix HA, Kalma JD (1972) Climate as a dominant control in the biogeography of northern Australia and New Guinea. In: Walker D (ed) Bridge and barrier: the natural and cultural history of Torres Strait. Australian National University, Canberra. pp 61–91

Philander SGH (1998) Learning from El Niño. Weather 53:270–274

Piper PJ, Ochoa J, Paz V, Lewis H, Ronquillo WP (2008) The first evidence for the past presence of the tiger *Panthera tigris* (L.) on the island of Palawan, Philippines: extinction in an island population. Palaeogeogr Palaeoclim Palaeoecol 264:123–127

Piper PJ, Ochoa J, Robles EC, Lewis H, Paz V (2010) Palaeozoology of Palawan Island, Philippines. Quatern Int. 223:142–158

Platnick NI (1992) Patterns of biodiversity. In: Eldredge N (ed) Systematics, ecology, and the biodiversity crisis. Columbia University Press, New York, NY. pp 15–24

Plaziat J-C, Cavagnetto C, Koeniguer J-C, Baltzer F (2001) History and biogeography of the mangrove ecosystem, based on a critical reassessment of the paleontological record. Wetlands Ecol Manage 9:161–179

Primack R, Corlett R (2005) Tropical rain forests. An ecological and biogeographical comparison. Blackwell, Oxford

Raes N, Roos MC, Slik JWF, Vanloon EE, Tersteege H (2009) Botanical richness and endemicity patterns of Borneo derived from species distribution models. Ecography 32:180–192

Randi E, Lucchini V, Armijo-Prewitt T, Kimball RT, Braun EL, Ligon JD (2000) Mitochondrial DNA phylogeny and speciation in the Tragopans. Auk 117:1007–1019

Richards PW (1998) The tropical rain forest: an ecological study, second edition. Cambridge University Press, Cambridge

Ridder-Numan JWA (1998) Historical biogeography of Spatholobus (Leguminosae-Papilionoideae) and allies in SE Asia. In: Hall R, Holloway JD (eds) Biogeography and geological evolution of SE Asia. Backhuys Publishers, Leiden

Roos MC, Keßler PJA, Gradstein SR, Baas P (2004) Species diversity and endemism of five major Malesian islands: diversity-area relationships. J Biogeogr 31:1893–1908

Sheue C-R, Liu H-Y, Tsai C-C, Rashid SMA, Yong JWH, Yang Y-P (2009) On the morphology and molecular basis of segregation of *Ceriops zippeliana* and *C. decandra* (Rhizophoraceae) from Asia. Blumea 54:220–227

Simons WJF, Socquet A, Vigny C, Ambrosius BAC, Haji Abu S, Promthong C, Subarya C, Sarsito DA, Matheussen S, Morgan P, Spakman W (2007) A decade of GPS in Southeast Asia. Resolving Sundaland motion and boundaries. J Geophys Res 112:B06420. doi:10.1029/2005JB003868

Stowe LL, Carey MR, Pellegrino PP (1992) Monitoring the Mt. Pinatubo aerosol layer with NOAA/11 AVHRR data. Geophys Res Lett 19:159–162

Stuijts I, Newsome JC, Flenley JR (1988) Evidence of Late Quaternary vegetation change in the Sumatran and Javan highlands. Rev Palaeobot Palynol 55:207–216

Sun X, Wang P (2005) How old is the Asian monsoon system?—Palaeobotanical records from China. Palaeogeogr Palaeoclim Palaeoecol 222:181–222

Sun X, Li X, Luo Y, Chen X (2000) The vegetation and climate at the last glaciation on the emerged continental shelf of the South China Sea. Palaeogeogr Palaeoclim Palaeoecol 160:301–316

Tjia HD (1980) The Sunda Shelf, Southeast Asia. Z Geomorphologie 24:405–427

Tosi AJ, Morales JC, Melnick DJ (2002) Y-chromosome and mitochondrial markers in *Macaca fasicularis* indicate introgression with Indochinese *M. mulatta* and a biogeographic barrier in the Isthmus of Kra. Int J Primatol 23:161–178

van den Bergh GD (1999) The Late Neogene elephantoid-bearing faunas of Indonesia and their palaeozoogeographic implications. A study of the terrestrial faunal succession of Sulawesi, Flores and Java, including evidence for early hominid dispersal east of Wallace's line. Scripta Geol 117:1–419

van den Bergh GD, de Vos J, Sondaar PY (2001) The late Quaternary palaeogeography of mammal evolution in the Indonesian archipelago. Palaeogeogr Palaeoclim Palaeoecol 171:385–408

van Steenis CGGJ (1950) The delimitation of Malaysia and its main plant geographical divisions. In: van Steenis (ed) Flora Malesiana Series I, vol 1. Noordhoff-Kolff n.v., Djakarta. pp lxx–lxxv

van Tol J (1994) The Odonata of Sulawesi and adjacent islands. Part 3. The genus *Macromia* Rambur (Corduliidae. Tijdschrift Entomol 137:87–94

van Welzen PC, Slik JWF (2009) Patterns in species richness and composition of plant families in the Malay Archipelago. Blumea 54:166–171

van Welzen PC, Slik JWF, Alahuhta J (2005) Plant distribution patterns and plate tectonics in Malesia. Biol Skrifter 55:199–217

Verstappen HT (1997) The effect of climatic change on southeast Asian geomorphology. J Quatern Sci 12:413–418

Voris HK (2000) Maps of Pleistocene sea levels in Southeast Asia: shorelines, river systems and time durations. J Biogeogr 27:1153–1167

Waits LP, Sullivan J, O'Brien SJ, Ward RH (1999) Rapid speciation events in the family Ursidae indicated by likelihood phylogenetic estimation from multiple fragments of mtDNA. Mol Phylogenet Evol 13:82–92

Walker D, Chen Y (1987) Palynological light on tropical rainforest dynamics. Quatern Sci Rev 6:77–92

Walker D, Flenley JR (1979) Late Quaternary vegetational history of the Enga Province of upland Papua New Guinea. Philos Trans R Soc Lond B 286:265–344

Wallace AR (1863) On the physical geography of the Malay archipelago. J Roy Geogr Soc 33: 217–234

Walsh RPD, Newbery DM (1999) The ecoclimatology of Danum, Sabah, in the context of the world's rainforest regions, with particular reference to dry periods and their impact. Philos Trans R Soc Lond B 354:1869–1883

Wang P (1999) Response of Western Pacific marginal seas to glacial cycles: paleoceanography and sedimentological features. Marine Geol 156:5–39

Whitmore TC (1981) Wallace's line and plate tectonics. Oxford monographs on biogeography 1. Clarendon Press, Oxford

Whitmore TC (1984) Tropical rain forests of the far east, 2nd edn. Clarendon Press, Oxford

Whitmore TC (1989) Southeast Asian tropical forests. In: Lieth BH, Werger MJA (eds) Ecosystems of the world 14. Elsevier, Amsterdam

Whitmore TC (1998) An introduction to tropical rain forests, 2nd edn. University Press, Oxford

Whitten AJ, Mustafa M, Henderson GS (1987) The ecology of Sulawesi. Gadjah Mada University Press, Yogyakarta

Whitten A, Soeriaatmadja RE, Suraya AA (1996) The ecology of Java and Bali. Periplus Editions, Singapore

Wich SA, Meijaard E, Marshall AJ, Husson S, Ancrenaz M, Lacy RC, van Schaik CP, Sugardjito J, Simorangkir T, Traylor-Holzer K, Doughty M, Supriatna J, Dennis R, Gumal M, Knott CD, Singleton I (2008) Distribution and conservation status of the orang-utan (*Pongo* spp) on Borneo and Sumatra: how many remain? Oryx 42:329–339

Willi Y, van Buskirk J, Hoffmann AA (2006) Limits to the adaptive potential of small populations. Annu Rev Ecol Evol Syst 37:433–458

Wilson MEJ, Moss SJ (1999) Cenozoic palaeogeographic evolution of Sulawesi and Borneo. Palaeogeogr Palaeoclim Palaeoecol 145:303–337

Wilting A, Buckley-Beason VA, Feldhaar H, Gadau J, O'Brien SJ, Linsenmair KE (2007) Clouded leopard phylogeny revisited. Support for species recognition and population division between Borneo and Sumatra. Front Zool 4:15. doi:10.1186/1742-9994-4-15

Wong KM (1998) Patterns of plant endemism and rarity in Borneo and the Malay Peninsula. In: Peng CI, Lowry PP (eds) Rare, threatened, and endangered floras of Asia and the Pacific Rim: Academia Sinica monograph series no. 16. Academia Sinica. Taipei. pp 139–169

Wong KM (2004) Bamboo, the amazing grass. A guide to the diversity and study of bamboos in Southeast Asia. International Plant Genetic Resources Institute, University of Malaya, Kuala Lumpur

Wong PP (2005) The coastal environment of Southeast Asia. In: Gupta A (ed) The physical geography of Southeast Asia. Oxford University Press, Oxford. pp 177–192

Wong KM, Low YW (2011) A revision of Philippine *Gardenia* (Rubiaceae). Edinb J Bot 68:11–32

Wong KM, Nilus R, Petol GH, Sugau JB, Pereira JT, Ong RC (1999) The landscape, vegetation and botany of the Semporna Islands of Sabah, Borneo: a conservation perspective. Sandakania 13:41–65

Wong KM, Sugumaran M, Lee DKP, Zahid MS (2010) Ecological aspects of endemic plant populations on Klang Gates quartz ridge, a habitat island in Peninsular Malaysia. Biodivers Conserv 19:435–447

Woodruff DS (1990) Genetics and demography in the conservation of biodiversity. J Sci Soc Thai 16:117–132

Woodruff DS (2010) Biogeography and conservation in Southeast Asia: how 2.7 million years of repeated environmental fluctuations affect today's patterns and the future of the remaining refugial-phase biodiversity. Biodivers Conserv 19:919–941

Woodruff DS, Turner LM (2009) The Indochinese-Sundaic zoogeographic transition: a description and analysis of terrestrial mammal species distributions. J Biogeogr 36:803–821

Wright SJ (2002) Plant diversity in tropical forests: a review of mechanisms of species coexistence. Oecologia 130:1–14

Wright SJ (2007) Plant diversity in tropical forests. In: Pugnaire FI, Valladares F (eds) Functional plant ecology. 2nd ed. CRC Press, Boca Raton, FL. pp 351–368

Wu M-J, Huang T-C, Huang S-F (2009) Phylogenetic biogeography of *Euphrasia* section *Malesiana* (Orobanchaceae) in Taiwan and Malesia. Blumea 54:242–247

Wurster CM, Bird MI, Bull ID, Creed F, Bryant C, Dungait JAJ, Paz V (2010) Forest contraction in north equatorial Southeast Asia during the Last Glacial Period. Proc Natl Acad Sci USA 107:15508–15511

Yap SY (2002) On the distribution patterns of Southeast-East Asian freshwater fish and their history. J Biogeogr 29:1187–1199

Zahid MS (2004) Nine novelties in *Porterandia* (Rubiaceae) from Borneo and Sulawesi. Sandakania 15:55–78

Zahid MS, Wong KM (2010) The circumscription, taxonomy and biogeography of *Porterandia* (Rubiaceae – Gardenieae). Edinb J Bot 67:265–342

Zhang Y, Chen Z, Shi L (1993) Phylogeny of the slow loris (genus *Nycticebus*): an approach using mitochondrial DNA restriction enzyme analysis. Int J Primatol 14:167–175

Zhang Y, Ryder OA, Zhang Y (2001) Genetic divergence on orangutan subspecies (*Pongo pygmaeus*). J Mol Evol 52:516–526

Chapter 3
Gene Flow, Mating Systems, and Inbreeding Depression in Natural Populations of Tropical Trees

Yoshihiko Tsumura

Abstract I have summarized and reviewed the data on mating system and gene flow of tropical tree species, especially dipterocarps. These are major canopy trees and important for forestry and ecosystem function in Southeast Asia. Microsatellite markers have been developed in many species of the Dipterocarpaceae to clarify the mating system, gene flow, and genetic structure. Dipterocarp species are a predominantly outcrossed species but the outcrossing rate is strongly influenced by the density of flowering tree, flowering phenology, and pollinator behavior. The pollen dispersal patterns of dipterocarp species were affected by differences in conspecific tree flowering density. Reductions in conspecific tree flowering density led to an increase in the selfing rate. In dipterocarp species, inbreeding depression can occur at all stages of development, including seed formation, seed germination, seedling growth, and sapling establishment. Genetic information discussed in this chapter may provide a more promising direction for conservation of natural forest and developing sustainable forestry.

Keywords Conservation · Dipterocarp · Flowering density · Microsatellite marker · Pollinator · Pollen flow · Southeast Asia

3.1 Introduction

The conservation of tropical rainforests is an important issue for biodiversity and has an impact on climate change. In Southeast Asia, forests have long been used for domestic purposes, such as house construction, fuel and food supply, and also as a source of medicinal plants. However, once large-scale commercial logging started, the extent of forests was dramatically reduced. To prevent the rapid decline of forests, selective logging systems have been used in a number of countries (Appanah 1988). Examples include the Selective Management System (SMS) in Peninsular

Y. Tsumura (✉)
Department of Forest Genetics, Forestry and Forest Products Research Institute, Tsukuba, Ibaraki 305-8687, Japan
e-mail: ytsumu@ffpri.affrc.go.jp

R. Wickneswari, C. Cannon (eds.), *Managing the Future of Southeast Asia's Valuable Tropical Rainforests*, Advances in Asian Human-Environmental Research 2,
DOI 10.1007/978-94-007-2175-3_3, © Springer Science+Business Media B.V. 2011

Malaysia, the Philippine Selective Logging System (PSLS), and the Indonesian Selective Cutting System, locally known as the TPI (Tebangan Pilih Indonesia). These systems are important for sustainable forestry. However, if more information was available for each system, such as number of remnant trees per hectare for each species after selective logging, based on genetic data, the high genetic diversity of important tree species might be maintained in logged forests.

Mating system and gene flow distance often differ between species. For conservation of populations and species, it is important to understand mating system and gene flow. Using such data, it may be possible to estimate the optimum mature tree density for healthy seed production for each species. Highly polymorphic DNA markers, such as microsatellite markers, allow us to estimate mating system and gene flow accurately. Collecting such data for tree species in tropical areas is essential, not only for a basic understanding of species' characteristics, but also for developing sustainable forestry and to support the conservation of natural forests.

In this chapter, I have summarized and reviewed the data on mating system and gene flow of tropical tree species, especially dipterocarps. These are major canopy trees and important for forestry and ecosystem function in Southeast Asia. Dipterocarp trees are predominantly allogamous species and thus inbreeding depression sometimes occurs when mating with relatives. Therefore, I also review inbreeding depression in dipterocarp species, as revealed by genetic data.

3.2 Microsatellite Markers

Genetic markers are useful for investigating mating system and gene flow in plant species. Allozyme markers have been used for a long time for the evaluation of mating system, such as outcrossing rate and biparental inbreeding, by means of the model developed by Ritland and Jain (1981). However, allozyme polymorphisms are not sufficiently abundant to assess gene flow directly through pollen and seed movement. At the beginning of the 1990s, highly polymorphic DNA markers, known as microsatellite markers, began to be used for gene flow studies. Microsatellites, also called simple sequence repeats (SSRs), are tandem repeats of short (1–6 nucleotide) motifs that are interspersed throughout eukaryotic genomes (Stallings et al. 1991; Charlesworth et al. 1994). The first microsatellite markers were developed in humans (Litt and Luty 1989; Weber and May 1989) and then in other mammalian and crop species. Microsatellite loci are co-dominant markers and characteristically exhibit high levels of variation in the number of repeat units. Furthermore, once specific primers have been designed, they can be amplified from small amounts of DNA by polymerase chain reaction (PCR). Due to these advantages, microsatellite markers are useful and powerful molecular tools for determining parents and for kinship analysis.

Microsatellite markers have been developed in many species of the Dipterocarpaceae, including *Dryobalanops lanceolata* (Terauchi 1994), *Shorea curtisii* (Ujino et al. 1998), *Neobalanocarpus heimii* (Iwata et al. 2000), *Shorea cordifolia* (Stacy et al. 2001), *Shorea leprosula* (Lee et al. 2004a), *Shorea lumutensis* (Lee et al. 2006), *Hopea bilitonensis* (Lee et al. 2004b), *Dryobalanops aromatica* (Nanami et al. 2007), and *Dipterocarpus tempehes* (Isagi et al. 2002). Microsatellite markers based on expressed sequence tags (ESTs) have been also developed in *S. leprosula* (Ng et al. 2009). A microsatellite marker can often be used with closely related species (Dayanandan et al. 1997; Ujino et al. 1998; Lee et al. 2004a). In particular, EST-SSR markers can be used with a relatively wide range of species, because the flanking region of EST-SSR markers is more conserved between species than that of genomic SSRs (Ng et al. 2009). Nevertheless, the polymorphism of EST-SSRs is generally lower than that of genomic SSRs.

3.3 Mating System

Plant species produce seeds by mating; briefly, the process is that pollen (♂) reaches the stigma, germinates, and the resultant pollen tube elongates down the stigma toward the ovule (♀), and fertilization occurs. Fertilization can occur with pollen from another individual or alternatively from the same flower or plant (selfing), giving rise to outcrossing and inbreeding, respectively. Different plant species can have different mating systems, which can also vary according to the environment and can evolve (Richards 1997). Pollen is transported by a number of vectors, including wind, water, insects, and birds. The outcrossing rate is affected by the kind of pollen vector and the density of mature trees.

Flowering phenology is also an important factor in mating success, especially for tropical tree species that occur at low densities in the canopy. The dipterocarp forests of Southeast Asia exhibit synchronized flowering, which is triggered (Ashton et al. 1988) by El Niño events. This can cause massive fruit production (Sakai et al. 1999). The degree of flowering synchrony with surrounding conspecific trees is another factor causing different outcrossing rates among trees. For example, individual *Shorea* trees have a relatively short flowering period of 16–26 days (Ashton et al. 1988).

Based on isozyme data, Murawski and Bawa (1994) reported that the outcrossing rate of the dipterocarp *Stemonoporus oblongifolius* was 84%. Kitamura et al. (1994) reported that the outcrossing rate of *D. aromatica*, another dipterocarp, ranged between 79.4 and 85.6% in primary and secondary forests, but was not significantly different between forest types. Using microsatellite markers, the average outcrossing rate of *S. curtisii* in primary forest was estimated to be 96.3% (Obayashi et al. 2002), which is within the range of estimates for tropical tree species reviewed by Loveless (1992) and those reported by other authors (Table 3.1). However, in a selectively logged plot, the average outcrossing rate was found to be only 52.2% (Obayashi et al. 2002), which is about half the average value in the undisturbed plot and also around half of the values reported in previous studies. Reductions of outcrossing rate in logged forests compared with primary forests have been reported for several tropical tree species: *Shorea megistophylla* (Murawski et al.

Table 3.1 Outcrossing rate and gene flow of dipterocarps and the relationship with forest type and main pollinator

Species	Forest type	Main pollinator	Materials	Outcrossing rate (%)	Mature tree density per hectare	Average gene flow distance (m)	Reference
Stemonoporus oblongifolius	Primary	Small insects	Seed	84.48	–	–	Murawski and Bawa (1994)
Dryobalanops aromatica	Primary	*Apis*, *Trigona*	Seed	85.60	–	–	Kitamura et al. (1994)
	Logged		Seed	79.40	–	–	
D. aromatica	Primary	*Apis*, *Trigona*	Seed	92.30	–	–	Lee (2000)
	Logged		Seed	76.60	–	–	
Dipterocapus tempehes	Primary	*Apis*	Seedling	93.0, 96.0	3.95	192, 222	Kenta et al. (2004)
Neobalanocarpus heimii	Primary	*Apis*, *Trigona*	Seed	87.70	–	–	Naito et al. (2005)
	Primary	*Apis*, *Trigona*	Seedling	93.07	0.71	191.2	Konuma et al. (2000)
	Primary		Sapling	100.00	–	–	
Shorea congestiflora	Logged	Small insects	Seed	87.40	6.40	–	Murawski et al. (1994)
Shorea megistophylla	Primary	Small insects	Seed	86.60	–	–	Murawski et al. (1994)
	Logged		Seed	71.30	–	–	
Shorea trapezifolia	Logged	Small insects	Seed	54.4, 61.7	24.00	–	Murawski et al. (1994)
Shorea acuminata	Primary	Small beetles	Seed	76.50	1.38	133	Naito et al. (2008)
Shorea leprosula	Primary	Small beetles, thrip	Seed	84.00	–	–	Nagamitsu et al. (2001)
S. leprosula	Primary	Small beetles, thrip	Seed	83.70	–	–	Lee et al. (2000)
S. leprosula	Primary	Small beetles, thrip	Seedling	86.85	0.55	369	Fukue et al. (2007)
S. leprosula	Primary	Small beetles, thrip	Seed	72.1, 74.5	1.52	60, 154	Tani et al. (2009)
Shorea parvifolia	Primary	Small beetles, thrip	Seed	51.9, 49.1	1.75	166, 137	Tani et al. (2009)
Shorea curtisii	Primary	Small beetles, thrip	Seed	96.30	–	–	Obayashi et al. (2002)
	Logged		Seed	52.20	–	–	

1994), *Pithecellobium elegans* (Hall et al. 1996), *Carapa procera* (Doligez and Joly 1997) and *D. aromatica* (Lee 2000). Lee (2000) documented that the outcrossing rate for *D. aromatica* was greater in primary forest than in secondary forest in Peninsular Malaysia. In contrast, Kitamura et al. (1994) found no significant difference between two types of forests in Brunei. Lee (2000) discussed the reasons for this difference between the regions and speculated that local microclimate and the microenvironment for pollinators might be different in the two areas. The high inbreeding rate in the selectively logged plot suggested that *D. aromatica* might not have a self-incompatibility mechanism, although some dipterocarp species do have such systems (Soepadmo 1989).

Outcrossing rates are strongly influenced by flowering tree density (Murawski and Hamrick 1992; Fukue et al. 2007) and the types and behavior of pollinators, which govern pollen movement (Ghazoul et al. 1998). Both of these factors may be partly responsible for the low outcrossing rate of *S. curtisii* observed in the selectively logged site. Tree density at the site was much lower than at the undisturbed site, and the principal pollinators for the genus *Shorea* are thought to be thrips and/or small beetles with low mobility (Appanah and Chan 1981; Ashton et al. 1988; Momose et al. 1998; Sakai et al. 1999). The outcrossing rate may not be strongly influenced by tree density if the species has a wide-ranging pollinator, such as *Apis* or *Trigona* (Kitamura et al. 1994; Konuma et al. 2000). However, as the outcrossing rate of *D. aromatica*, which has high-mobility pollinators such as *Apis*, was reduced by selective logging, the microclimate and microenvironment may also influence pollinator behavior (Lee 2000).

The outcrossing rate varied greatly among individual trees of *S. curtisii* within both plots, but especially in the selectively logged plot (Obayashi et al. 2002). The outcrossing rate of individuals should depend on the neighborhood density of conspecific flowering trees and be lower in isolated individuals than in clusters of trees (Murawski et al. 1990). Nagamitsu et al. (2001) also suggested that *S. leprosula*, a tree species with a low outcrossing rate and a similar pollinator, might be isolated from reproductive neighbors in the Pasoh forest reserve. Fukue et al. (2007) reported that selfing rate of each tree of *S. leprosula* was related to the density of surrounding mature trees. The degree of flowering synchrony with surrounding conspecific trees could also cause variations in outcrossing rate among these trees, since individual *Shorea* trees have relatively short flowering periods, as mentioned earlier (Ashton et al. 1988).

The number of alleles per locus (*N*a) and heterozygosity (*H*e) of the adult *S. curtisii* population in the selectively logged plot, in which 30% of the wood volume was removed and the density of mature trees was decreased by approximately 75%, showed similar values to those from the undisturbed plot (Obayashi et al. 2002). This suggests that the disturbance caused by selective logging exerts little influence on the population genetic parameters *N*a and *H*e, as evaluated by highly variable microsatellite markers. Expected heterozygosity, *H*e, is less sensitive to population bottlenecks than the proportion of polymorphic loci or the number of alleles per locus (Barrett and Kohn 1991). In contrast, the mean number of alleles per locus per maternal tree derived from paternal trees by outcrossing was significantly lower in the selectively logged *S. curtisii* plot, compared with the undisturbed plot (Obayashi et al. 2002). Disturbances such as selective logging are thought to cause a strong bottleneck and thus reduce the effective population size in the following generation. This suggests that some degree of genetic variation might be lost in the next generation of selectively logged forest trees that germinate from newly produced seeds. This is because the total number of alleles per unit area of the future generation will be decreased if new alleles are not supplied through pollen flow and/or seed dispersal from outside the plot. Consistent with this hypothesis, Dayanandan et al. (1999) reported that allelic richness was lower in the sapling cohort of an isolated fragment of *Carapa guianensis* forest, compared with a more extensive stand. In allogamous species, such as dipterocarps, the low outcrossing rate may influence the genetic variation of the future forest. Thus, knowledge of pollinator behavior and flowering tree density for each species must be obtained. In order to evaluate differences in the genetic diversity between forests with different logging histories, parameters such as allelic richness and rare allele richness are better than heterozygosity.

3.4 Gene Flow Through Pollen and Seed

Genes can be moved through pollen and seeds. In general, pollen movement by vectors such as wild animals, insects, and birds occurs over greater distances than seeds, though seeds can be moved by birds and mammals over significant distances. These long distance movements of pollen and seed are important for maintaining genetic diversity and reducing inbreeding depression.

Relatively long distance pollen flow among flowering trees was observed, using microsatellite markers, for *N. heimii* in a 36 ha study plot in Pasoh (Konuma et al. 2000). Pollen from outside the study plot sired 35% of the offspring in the study, and the estimated average mating distance was 524 m (Table 3.1). *N. heimii* is mainly pollinated by stingless bees or honeybees (Appanah 1985), which are known to be long distance pollinators (Dayanandan et al. 1990).

Fukue et al. (2007) demonstrated that flowering tree density affected gene flow of *S. leprosula* in a 100 ha research plot in the Ampang forest reserve. In a population with high mature tree density, pollination distances were frequently within 200 m, but in populations with low adult tree density the distances were longer. The selfing rate tended to be higher in the population with low mature tree density. These findings suggest that the density of flowering trees affects selfing rates, gene flow, and, thus, the genetic diversity of *S. leprosula* populations.

Tani et al. (2009) also suggested that the pollen dispersal patterns of two dipterocarp species, *S. leprosula* and *Shorea parvifolia*, were affected by differences in conspecific tree flowering density. Reductions in conspecific tree flowering density led to an increase in the selfing rate. Active pollen dispersal and a larger number of effective paternal parents were observed for both the species in the larger magnitude of general flowering season.

Long distance pollen flow most probably depends on the performance of pollinators, and pollinator behavior is probably affected by the distance between flowering trees (Bawa 1998). Thus, flowering tree density can also influence the effective mating distance of tropical trees. For example, Ghazoul et al. (1998) reported that a bee pollinator of *Shorea siamensis* flew much further at a site where there were long distances between flowering host trees than it did at sites where the distances between flowering host trees were shorter. However, there must be limits on the movements of pollinators between flowering trees.

Wind dispersal of seeds is more frequently found in canopy and emergent trees than among trees of the forest understory (Turner 2001). Using microsatellite markers, Takeuchi et al. (2004) reported the differences in seed dispersal ability between three dipterocarp species, *Hopea dryobalanoides*, *S. parvifolia* and *Shorea acuminata*. Differences were also found among these species in their range of physical seed dispersal, which depended on the height of the seed release point. Taller trees dispersed their seed more extensively. The estimates of seed dispersal area were consistent with the degree of spatial genetic structure found in the three species.

Seed migration patterns of *N. heimii* have also been inferred from the identification and location of offspring with haplotypes that were incompatible with those of neighboring reproductive trees (Konuma et al. 2000). Clearly, fruits or seeds of *N. heimii* had been carried over significant distances by a dispersal vector, possibly small mammals. In tropical forests, squirrels disperse fruits of some tree species (Becker et al. 1995). Nagamitsu et al. (2001) and Fukue et al. (2007) reported similar phenomena in *S. leprosula* using microsatellite analysis, which was second dispersal of seeds from the mother trees.

Thus, microsatellite analysis is a highly sensitive method for determining gene flow within forests, and it can be used to identify the parents of seeds, seedlings, and saplings at a study site. Such analyses will provide us with an understanding of the range and scope of gene flow, including pollen flow and seed dispersal parameters, for important forest tree species and thus provide essential information to help design sustainable forestry regimes.

3.5 Inbreeding Depression

If seeds are produced by consanguineous matings in allogamous plants, inbreeding depression may be manifested at many subsequent developmental stages, such as fruit maturation (Nagamitsu et al. 2001; Naito et al. 2005, 2008), germination (Naito et al. 2005, 2008), seedling growth (Konuma et al. 2000), or sapling development. Dipterocarps are predominantly outcrossing species, so inbreeding depression may be expected only in selfed seeds. In *S. leprosula*, the inbreeding rate was reduced in mature seeds compared with immature seeds from one mother tree which suggested that inbreeding depression and/or self-incompatibility might occur during fruit maturation (Nagamitsu et al. 2001).

Mean seed mass of selfed progeny was lower than that of outcrossed progeny in the tree *N. heimii*. Since smaller seeds suffered a fitness disadvantage at germination, the reduced seed mass of selfed progeny would be one of the determinants of the observed inbreeding depression during seedling establishment (Naito et al. 2005). Inbreeding depression during seedling establishment was manifested as an overall decrease in the survival of selfed progeny from seeds to established seedlings. Since there were no significant differences in the proportion of selfed progeny between immature and mature seed stages in another species, *S. acuminata*, there was no evidence of selective abortion of selfed seeds during seed development (Naito et al. 2008). However, the seed mass of outcrossed progeny was heavier than that of selfed progeny, and heavier seeds showed higher success rates at germination and seedling establishment, as was reported for *N. heimii*.

Konuma et al. (2000) found selfed offspring among seedlings but not saplings of *N. heimii* and suggested that inbreeding depression may occur via selection of surviving seedlings. Self-incompatibility effectively avoids inbreeding, and some dipterocarp species are known to utilize this system (Soepadmo 1989). However, in low-density populations caused by selective logging, the probability of mating via foreign pollen might decline if the main pollinator does not disperse far. Lee (2000), in a study of *D. aromatica*, suggested that selective logging might increase inbreeding by consanguineous mating. He compared the values of the adjusted outcrossing rate (mutilocus outcrossing rate (tm) – single locus outcrossing rate (ts)) between primary and logged forests and found the (tm – ts) value of primary forest was higher than that of logged forest, which suggested biparental inbreeding. Thus, we have to maintain a suitable flowering tree density and environment for pollinators to ensure genetically sustainable forestry, according to the biological and ecological characteristics of the trees concerned.

3.6 Conclusions and Perspectives

The outcrossing rate of dipterocarp species is generally high. However, in forests with a lower density of mature trees, the outcrossing rate tends to be low, depending on the main pollinators. Poorly dispersing pollinators will result in lower outcrossing, especially where mature trees are at a low density. Species with self-incompatibility systems do not self-pollinate, but other species will have increasing proportions of selfed seeds as the density of mature or flowering trees decreases. For conservation and sustainable forestry, an appropriate tree density for each species should be maintained, even in selectively logged forests. Genetic data should be used to determine the optimum tree density.

Gene flow through pollen is also dependent on the main pollinator: strong flyers, such as *Apis* and *Trigona*, can carry pollen over longer distances than weak flyers, such as small beetles and thrips. Most seed dispersal is by wind, with small mammals acting as seed vectors for secondary dispersal. In tropical areas, most tree species have an animal pollination system. Thus, species and ecosystem diversity are essential for successful reproduction of tree species.

Predominantly allogamous tree species show severe genetic depression when inbreeding occurs. In dipterocarp species, such depression can occur at all stages of development, including seed formation, seed germination, seedling growth, and sapling establishment.

Southeast Asia has large areas of tropical forests, although these have declined dramatically as a result of commercial logging. New priorities are to develop sustainable forestry and to conserve important natural forests. Selective logging has been conducted in these areas, but there is no conclusive evidence that the current selective system results in sustainable forestry. Genetic information discussed in this chapter may provide a more promising direction for developing sustainable forestry. If we can provide the minimum number of remnant trees of each species after selective logging, based on the genetic data, genetic diversity can be maintained and inbreeding depression avoided. Mating and gene flow patterns tend to be similar in species with similar ecological characteristics. Therefore, information on the mating system, gene flow, and inbreeding depression relating to major species could be relevant to closely related species and be used to determine the minimum number of mother trees to be retained for successful breeding. Of course, the main pollinators, flowering phenology, and synchrony of individual species are also important data when determining the minimum number of trees to retain.

Acknowledgments The author would like to thank Naoki Tani for providing the data on gene flow distances of *S. leprosula* and *S. parvifolia*. The study was partly funded by the Global Environment Research Program supported by the Ministry of Environment in Japan (grant no. E-091) and a Grant-in-Aid for Scientific Research (No. 18255010) provided by the Ministry of Education, Culture, Sports, Science and Technology of Japan.

References

Appanah S (1985) General flowering on the climax rain forests of South-east Asia. J Trop Ecol 1:225–240

Appanah S (1988) Management of natural forests. In: Appanah S, Turnbull JM (eds) A review of dipterocarps: taxonomy, ecology and silviculture. CIFOR, Bogor, Indonesia, pp 133–149

Appanah S, Chan HT (1981) Thrips: the pollinators of some dipterocarps. Malays Forester 44: 234–252

Ashton PS, Givnish TJ, Appanah S (1988) Staggered flowering in the Dipterocarpaceae: new insights into floral induction and the evolution of mast fruiting in the seasonal tropics. Am Nat 132:44–66

Barrett SH, Kohn JR (1991) Genetic and evolutionary consequences of small population size in plants: implications for conservation. In: Falk DA, Holsinger KE (eds) Genetics and conservation of rare plants. Oxford University Press, New York, NY, pp 3–30

Bawa K (1998) Conservation of genetic resources in the Dipterocarpaceae. In: Appanah A, Turnbull JM (eds) A review of dipterocarps: taxonomy, ecology and silviculture. Center for International Forestry Research, Bogor, Indonesia, pp 45–55

Becker P, Leighton M, Payne JB (1995) Why tropical squirrels carry seeds out of source crowns. J Trop Ecol 1:183–186

Charlesworth B, Sniegowaski P, Stephan W (1994) The evolutionary dynamic of repetitive DNA in eukaryotes. Nature 371:215–220

Dayanandan S, Attygalla DNC, Abeygunaskera AWWL, Gunatilleke IAUN, Gunatilleke CVS (1990) Phenology and floral morphology in relation to pollination of some Sri Lankan dipterocarps. In: Bawa KS, Hadley M (eds) Reproductive ecology of tropical forest plants Man and the biosphere series, vol 7. UNESCO, Paris, and Parthenon Publishing Group, Carnforth, pp 103–133

Dayanandan S, Bawa KS, Kesseli R (1997) Conservation of microsatellite among tropical trees (Leguminosae). Am J Bot 84:1658–1663

Dayanandan S, Dole J, Bawa K, Kesseli R (1999) Population structure delineated with microsatellite markers in fragmented populations of a tropical tree, *Carapa guianensis* (Meliaceae). Mol Ecol 8:1585–1592

Doligez A, Joly HI (1997) Mating system of *Carapa procera* (Meliaceae) in the French Guiana tropical forest. Am J Bot 84:461–470

Fukue Y, Kado T, Lee SL, Ng KKS, Muhammad N, Tsumura Y (2007) Effects of flowering tree-density on mating system and gene flow in *Shorea leprosula* (Dipterocarpaceae) in Peninsular Malaysia. J Plant Res 120:413–420

Ghazoul J, Liston KA, Boyle TJB (1998) Disturbance-induced density-dependent seed set in *Shorea siamensis* (Dipterocarpaceae), a tropical forest tree. J Ecol 86:462–473

Hall P, Walker S, Bawa KS (1996) Effect of forest fragmentation on genetic diversity and mating system in tropical tree, *Pithecellobium elegans*. Conserv Biol 10:757–768

Isagi Y, Kenta T, Nakashizuka T (2002) Microsatellite loci for a tropical emergent tree, *Dipterocarpus tempehes* V. Sl. (Dipterocarpaceae). Mol Ecol Notes 2:12–13

Iwata H, Konuma A, Tsumura Y (2000) Development of microsatellite markers in the tropical tree *Neobalanocarpus heimii* (Dipterocarpaceae). Mol Ecol 9:1684–1685

Kenta T, Isagi Y, Nakagawa M, Yamashita M, Nakashizuka T (2004) Variation in pollen dispersal between years with different pollination conditions in a tropical emergent tree. Mol Ecol 13:3575–3584

Kitamura K, Rahman MYBA, Ochiai Y, Yoshimaru H (1994) Estimation of the outcrossing rate on *Dryobalanops aromatica* Gaertn. f. in primary and secondary forests in Brunei, Borneo, Southeast Asia. Plant Species Biol 9:37–41

Konuma A, Tsumura Y, Lee CT, Lee SL, Okuda T (2000) Estimation of gene flow in the tropical-rainforest tree *Neobalanocarpus heimii* (Dipterocarpaceae), inferred from paternity analysis. Mol Ecol 9:1843–1852

Lee SL (2000) Mating system parameters of *Dryobalanops aromatica* Gaertn. F. (Dipterocarpaceae) in three different forest types and a seed orchard. Heredity 85:338–345

Lee SL, Wickneswari R, Mahani MC, Zakari AH (2000) Mating system parameters in a tropical tree species, *Shorea leprosula* Miq. (Dipterocarpaceae) from Malaysian lowland dipterocarp forest. Biotropica 32:693–702

Lee SL, Tani N, Ng KKS, Tsumura Y (2004a) Characterization of 15 polymorphic microsatellite loci in an endangered tropical tree *Hopea bilitonensis* (Dipterocarpaceae) in Peninsular Malaysia. Mol Ecol Notes 4:147–149

Lee SL, Tani N, Ng KKS, Tsumura Y (2004b) Isolation and characterization of 21 microsatellite loci in an important tropical tree *Shorea leprosula* and their applicability to *S. parvifolia*. Mol Ecol Notes 4:222–225

Lee SL, Ng KKS, Sawa LG, Lee CT, Muhammada N, Tani N, Tsumura Y, Koskela J (2006) Linking the gaps between conservation research and conservation management of rare dipterocarps: a case study of *Shorea lumutensis*. Biol Conserv 131:72–92

Litt M, Luty JA (1989) A hypervariable microsatellite revealed by in vitro amplification of a dinucleotide repeat within the cardiac muscle actin gene. Am J Hum Genet 44:397–401

Loveless MD (1992) Isozyme variation in tropical trees: patterns of genetic organization. In: Adams WT, Strauss SH, Copes DL, Griffin AR (eds) Population genetics of forest trees. Kluwer, The Netherlands, pp 67–94

Momose K, Yumoto T, Nagamitsu T, Kato M, Nagamasu H, Sakai S, Harrison RD, Itioka T, Hamid AA, Inoue T (1998) Pollination biology in a lowland dipterocarp forest in Sarawak, Malaysia.

I. Characteristics of the plant-pollinator community in a lowland dipterocarp forest. Am J Bot 85:1477–1501

Murawski DA, Bawa KS (1994) Genetic structure and mating system of *Stemonoporus oblongifolius* (Dipterocarpaceae) in Sri Lanka. Am J Bot 81:155–160

Murawski DA, Hamrick JL (1991) The effect of the density of flowering individuals on the mating systems of nine tropical tree species. Heredity 67:167–174

Murawski DA, Hamrick JL (1992) The mating system of *Cavanillesia platanifolia* under extremes of flowering-tree density: a test of predictions. Biotropica 24:99–101

Murawski DA, Hamrick JL, Hubell SP, Foster RB (1990) Mating systems of two Bombacaceous trees of a neotropical moist forest. Oecologia 82:501–506

Murawski DA, Gunatilleke IAUN, Bawa KS (1994) The effect of selective logging on inbreeding in *Shorea megistophylla* (Dipterocarpaceae) from Sri Lanka. Conserv Biol 8:997–1002

Nagamitsu T, Ichikawa S, Ozawa M, Shimamura R, Kachi N, Tsumura Y, Muhammad N (2001) Microsatellite analysis of the breeding system and seed dispersal in *Shorea leprosula* (Dipterocarpaceae). Int J Plant Sci 162:155–159

Naito Y, Konuma A, Iwata H, Suyama Y, Seiwa K, Okuda T, Lee SL, Norwati M, Tsumura Y (2005) Selfing and inbreeding depression in seeds and seedlings of *Neobalanocarpus heimii*. (Dipterocarpaceae). J Plant Res 118:423–430

Naito Y, Kanzaki M, Iwata H, Obayashi K, Lee SL, Muhammad N, Okuda T, Tsumura Y (2008) Density-dependent selfing and its effects on seed performance in a tropical canopy tree species, *Shorea acuminata* (Dipterocarpaceae). Forest Ecol Manage 256:375–383

Nanami S, Ikeda S, Tani N, Tan S, Diway B, Harada K, Tsumura Y, Itoh A, Yamakura T (2007) Development of microsatellite markers for *Dryobalanops aromatica* (Dipterocarpaceae), a tropical emergent tree in Southeast Asia. Mol Ecol Notes 7:623–625

Ng KKS, Lee SL, Tsumura Y, Ueno S, Ng CH, Lee CT (2009) EST-SSRs isolated from *Shorea leprosula* and their transferability to 36 species within the Dipterocarpaceae. Mol Ecol Resour 9:393–398

Obayashi K, Tsumura Y, Ihara-Ujino T, Niiyama K, Tanouchi H, Suyama Y, Washitani I, Lee CT, Lee SL, Muhammad N (2002) Genetic diversity and outcrossing rate between undisturbed and selectively logged forests of *Shorea curtisii* (Dipterocarpaceae) using microsatellite DNA analysis. Int J Plant Sci 163:151–158

Richards AJ (1997) Plant breeding system, 2nd edn. Chapman & Hall, London

Ritland K, Jain SK (1981) A model of for the estimation of outcrossing rate and gene frequencies using n independent loci. Heredity 47:35–52

Sakai S, Momose K, Yumotoa T, Kato M, Inoue T (1999) Beetle pollination of Shorea parvifolia (section Mutica, Dipterocarpaceae) in a general flowering period in Sarawak, Malaysia. Am J Bot 86:62–69

Soepadmo E (1989) Contribution of reproductive biological studies towards the conservation and development of Malaysian plant genetic resources. In: Zakri AH (ed) Genetic resources of under-utilised plants in Malaysia. Malaysian National Committee on Plant Genetic Resources, Kuala Lumpur, pp 1–41

Stacy EA, Dayanandan S, Dancik BP, Khasa PD (2001) Microsatellite DNA markers for the Sri Lankan rainforest tree species, *Shorea cordifolia* (Dipterocarpaceae), and cross-species amplification in *S. megistophylla*. Mol Ecol Notes 1:53–54

Stallings RL, Ford AF, Nelson D, Torney DC, Hildebrand CE, Moyzis RK (1991) Evolution and distribution of (GT)n repetitive sequences in mammalian genomes. Genomics 10:807–815

Takeuchi Y, Ichikawa S, Konuma A, Tomaru N, Niiyama K, Lee SL, Muhammad N, Tsumura Y (2004) Comparison of the fine-scale genetic structure of three dipterocarp species. Heredity 92:323–328

Tani N, Tsumura Y, Kado T, Taguchi Y, Lee SL, Muhammad N, Ng KKS, Numata S, Nishimura S, Konuma A, Okuda T (2009) Paternity analysis-based inference of pollen dispersal patterns, male fecundity variation and influence of flowering tree densities in two dipterocarp species. Ann Bot. doi:10.1093/aob/mcp252

Terauchi R (1994) A polymorphic microsatellite marker from the tropical tree *Dryobalanops lanceolata* (Dipterocarpaceae). Jpn J Genet 69:567–576

Turner IM (2001) The ecology of trees in the tropical rain forest. Cambridge University Press, Cambridge

Ujino T, Kawahara T, Tsumura Y, Nagamitsu T, Wickneswari R, Yoshimaru H (1998) Development and polymorphism of simple sequence repeat DNA markers for *Shorea curtisii* and other Dipterocarpaceae species. Heredity 81:422–428

Weber JL, May PE (1989) Abundant class of human DNA polymorphisms which can be typed using the polymerase chain reaction. Am J Hum Genet 44:388–396

Chapter 4
Threats to Genetic Viability of Southeast Asian Forest Species

Ratnam Wickneswari

Abstract The major threat to genetic viability of the Southeast Asian forest species is commercial logging and fragmentation being largely due to infrastructural development and agricultural activities. Species vulnerability to the threat of genetic viability posed by commercial logging is highly correlated with its abundance in a particular forest management unit. Tree density for the species can be a useful indicator in reflecting the risk of genetic viability rather than the overall disturbance level based on reduction in basal area of all trees. Reproductive success of mature trees together with successful regeneration of seedlings and saplings in disturbed forests is an important determinant of genetic viability of species. Knowledge of the biological attributes of species including their local evolution can be used to develop field guides in harvesting operations to ensure the genetic quality of residuals left behind in the post-harvest stands.

Keywords Genetic viability · Species vulnerability · Tree density · Reproductive success · Commercial logging · Fragmentation

4.1 Introduction

In many countries in tropical Asia and the Pacific, both the rapid loss and the degradation of forests and associated genetic resources are posing a serious problem. Asia and the Pacific region account for about 70% of the world production of tropical industrial roundwood and harbour a rich and diverse array of plants, animals and microorganisms. The majority of timber species harvested are indigenous species from natural stands. According to FAO, currently 40% of tropical forests have been lost, with tropical Asia and the Pacific recording the highest harvesting rate.

Several steps have been taken to alleviate the problem including reforestation programs and designating forest reserves and national parks in which felling is limited or prohibited. Some efforts have been made by national and international research organisations, individually and collectively, in basic studies of genetic

---

R. Wickneswari (✉)
Faculty of Science and Technology, School of Environmental and Natural Resource Sciences, Universiti Kebangsaan Malaysia, 43600 Bangi, Selangor Darul Ehsan, Malaysia
e-mail: wicki@ukm.my

R. Wickneswari, C. Cannon (eds.), *Managing the Future of Southeast Asia's Valuable Tropical Rainforests*, Advances in Asian Human-Environmental Research 2,
DOI 10.1007/978-94-007-2175-3_4, © Springer Science+Business Media B.V. 2011

conservation such as reproductive biology, genetic variation of important tree species, ecology of tropical rainforests and the development of management systems for natural forests.

In spite of these efforts and warnings, the extent and quality of populations of indigenous commercial species in the tropical production forests (which now occupy a large part of the remaining tropical rainforests) are still diminishing. The timber of these species, especially members of the Dipterocarpaceae and Leguminosae, is particularly important for timber production. The decrease and degradation of their genetic resources will make it difficult to secure the breeding populations and planting stocks for sustained production of the valuable timber. Harvesting practices that do not result in the retention of superior phenotypes as seed trees after a logging operation may further aggravate the loss of superior genotypes of valuable timber species.

Assessment of genetic diversity can assist in identifying the patterns of variation in genetic diversity for tree improvement and genetic conservation. Maintenance of genetic diversity in forest tree populations that are undergoing population changes due to natural or human-induced events is seen to be instrumental for population viability, adaptation and continued evolution (Müller-Starck 1985; Gregorius et al. 1985; Ledig 1988; Namkoong 1991; Müller-Starck et al. 1992; Bush and Smouse 1992; Templeton 1995).

4.2 Genetic Diversity

Traditionally, genetic variation of tree species has been understood through assessment of survival and growth performance parameters in common garden-type trials such as species, provenance and progeny tests. Notwithstanding the long-term benefits of such activities in terms of availability of plant materials for continuous assessments and evaluation of genotype × environment interactions, these exercises can be time consuming, labour intensive and costly. Advances in molecular biology and biotechnology offer quick detection of genetic variation and characterisation of genotypes using molecular markers. Molecular markers include isozymes, restriction fragment length polymorphisms (RFLPs), random amplified polymorphic DNAs (RAPDs), directed amplification of minisatellite DNA (DAMD), amplified fragment length polymorphisms (AFLPs), simple sequence repeats (SSR) and single nucleotide polymorphisms (SNPs). These markers can be effectively used to provide genetic information that is complementary to forest inventory data and regeneration status for sustainable management of forests besides being employed for designing conservation strategies and breeding programmes.

Hamrick et al. (1979) who surveyed various life history traits found that species with large ranges, high fecundities, an outcrossing mode of reproduction, wind pollination, long generation time and from habitats representing later stages of succession have more genetic variation. Brown (1979) showed that patterns of interpopulation differentiation and heterozygosity within populations vary due to many complex factors affecting mating systems. Partial inbreeding due to restricted numbers of reproductive individuals or neighbourhood size is important to many outbreeding species.

In the last two decades, molecular markers have been employed to understand the genetic structure of natural populations of several tree species in Southeast Asia's tropical rainforests. These include comprehensive studies on *Pinus kesiya* (Boyle et al. 1990), *Pinus merkusii* (Szmidt et al. 1996), *Pterocarpus macrocarpus* (Liengsiri et al. 1995), *Stemonoporus oblongifolius* (Murawski and Bawa 1994), *Hopea odorata* (Wickneswari et al. 1995), *Tectona grandis* (Kertadikara and Prat 1995; Changtragoon and Szmidt 2000), *Shorea macrophylla* (Kanzaki et al. 1996), *Shorea leprosula* (Lee et al. 2000a; 2001; Cao et al. 2006), *Dryobalanops aromatica* (Lee et al. 2000b; Lim et al. 2002), *Hopea bilitonsis* (Lee et al. 2004), *Shorea parvifolia* (Cao et al. 2006), *Neobalanocarpus heimii* (Lee 2007), *Intsia palembanica* (Lee et al. 2002a), *Shorea lumutensis* (Lee et al. 2006), *Koompassia malaccensis* (Lee et al. 2007a; 2008), and several *Shorea* species (Cao et al. 2009). Most studies have focused on species of not only with known economic value but also subject to current or projected threats. Examples of such studies in Indonesia, Malaysia and Thailand are given in Boxes 4.1, 4.2 and 4.3, respectively.

Box 4.1 Genetic Diversity Studies of Dipterocarpaceae in Indonesia – *Shorea* Species

Shorea is largest genus of the Dipterocarpaceae and the most important tree family in tropical Southeast Asia both from an ecological and from an economic perspective. It encompasses about 200 species, of which 163 are distributed in Malesia, mostly in Indonesia, in particular on Sumatra and Borneo (Kalimantan) (Ashton 1982). The genetic diversity and structure of nine *Shorea* species from two different locations in Indonesia viz. Nanjak Makmur in Sumatra and Sumalindo in Borneo were evaluated using amplified fragment length polymorphism (AFLP) markers by Cao et al. (2009). Levels of genetic diversity of these species ranged from $H_e = 0.100$ for *Shorea acuminata* to $H_e = 0.165$ for *Shorea blumutensis*. The population of *S. blumutensis*, a rare species, possessed the highest genetic diversity suggesting that geographically restricted species can have levels of genetic variation comparable to closely related widespread common congeners. Analyses of molecular variance revealed that the genetic variation was mainly found among species in both locations (57.7% in Sumatra and 56.3% in Borneo).

All the individuals were clearly divided into two wood varieties, Red Meranti and Yellow Meranti. Within Red Meranti, the separation of sections (Brachypterae and Mutica) was not clear. *Shorea macroptera* from section Mutica (subsection Auriculatae) grouped together with species of section Brachypterae. The possibility of interspecific hybridisation among sympatrically occurring *Shorea* species was inferred from the genetic distance estimates. A combination of species-specific markers is required to discount any rare hybridisation event or shared ancestral variation.

Box 4.2 Genetic Diversity Studies of Leguminosae in Malaysia – *Intsia palembanica and Koompassia malaccensis*

Intsia palembanica is an important and widespread timber species in Malaysia. Because the species is known to be more valuable than others, when subjected to a greater degree of exploitation, it is expected that the species will be more vulnerable to become endangered. Locally known as Merbau, it occurs from the Andaman Islands, Thailand and Malesia eastwards to western New Guinea in inland lowland forest up to 1000 m above sea level (asl).

Lee et al. (2002a) investigated allozyme variation in *I. palembanica* from six natural populations in Peninsular Malaysia. They showed that levels of genetic diversity of *I. palembanica* populations in Peninsular Malaysia were comparably high (mean number of alleles per polymorphic locus was 2.4, mean effective number of alleles per polymorphic locus was 1.64, and mean expected heterozygosity (H_e) was 0.242), with the majority of the diversity being partitioned within populations ($G_{ST} = 0.040$ and $F_{ST} = 0.048$).

Intsia palembanica exhibited a mixed-mating system, with multilocus outcrossing rate of 0.766. Probability tests of recent effective population size reduction using the infinite allele model showed the occurrence of genetic bottlenecks on two populations. A simulation study based on pooled samples circumscribed the effective population size to 200 and 210 individuals. This estimate should be considered as exploratory, preliminary information on the management of the species and the Virgin Jungle Reserves.

Koompassia malaccensis is another important and widespread tropical timber tree species in Malaysia. It is distributed in Sumatra, Peninsular Malaysia, Singapore and Borneo (Hou 2000). The species is locally known as kempas and grouped under medium hardwood. The genetic diversity of *K. malaccensis* was investigated using allozyme (six populations; Lee et al. 2007a) and microsatellite markers (19 populations; Lee et al. 2008). Overall, all the populations showed high levels of genetic diversity. Majority of the diversity was partitioned within-population ($F_{ST} = 0.045$ and $R_{ST} = 0.07$). No correlation was detected between geographical distance and genetic distance.

Box 4.3 Genetic Studies of Leguminosae and Pinaceae in Thailand – *Pterocarpus macrocarpus*, *Dalbergia cochinchinensis* and *Pinus kesiya*

Pterocarpus macrocarpus is a highly valued species yielding furniture quality timber. It is distributed throughout most of mainland Southeast Asia (Coles and Boyle 1999). Its high economic value has led to the development of plantation programmes in both Myanmar and Thailand and even towards the extremes of its range; for example, in Nghe An Province, Vietnam, local agencies have established propagule production plantations (Nguyen Hoang Nghia, personal communication). In support of such efforts, there have been various studies on methods to improve propagation (Saw 1984), seedling growth (Whuangplong et al. 1992) and inter- and intra-population variation in seed and fruit characteristics (Piewluang et al. 1997). These studies suggest that there are exploitable differences among populations and significant variation within populations. Populations from western and central Thailand perform better on test sites in the central part of the country.

In parallel to this research on quantitative variation, molecular markers have been used to examine levels of genetic diversity in natural populations across Thailand (Liengsiri et al. 1995). The results indicate a gradient in levels of genetic diversity and rates of inbreeding from diverse, outcrossing populations in the west to less diverse and more inbred populations in the east. Whether this is a real gradient or is simply correlated with the greater levels of degradation in forest fragments in the east will require wider geographic sampling to resolve.

There has been a similar series of parallel studies on *D. cochinchinensis* – a species with even more valuable wood, but facing more intense threats. Soonhuae and Sziklai (1994) describe quantitative variation among 85 open-pollinated families from seven populations, while Soonhuae et al. (1995) describe isozyme variation for a subset of the same populations. Very similar trends were found as those reported for *P. macrocarpus*.

Similarly, genetic diversity studies have been undertaken for *P. kesiya*, a tropical pine species that is distributed naturally in mainland Southeast Asia. Das and Stephan (1984) reported on growth rate differences among populations, while Boyle et al. (1990) studied isozyme variation among four populations in Thailand. Results suggest that in most respects, the pattern and level of genetic variation in *P. kesiya* is similar to related temperate species in the same genus.

4.3 Genetic Viability

Studies of genetic diversity can have several very distinct and different goals. Genetic diversity studies can be linked to the management goal of identifying high yielding genetic resources for use in production forestry – usually in the context of plantation forestry. At the same time, genetic diversity studies are essential not only for the design and monitoring of genetic conservation programmes but also for assessing genetic erosion and viability of natural forests due to threats viz. harvesting of wood, grazing, fire and harvesting of non-timber forest products (Namkoong et al. 1996; 2002). While the goals of such studies are very different, some of the methodologies used are similar or identical. For example, the identification of high-value genetic resources for use in production programmes may make use of field provenance and progeny trials, but tree improvement can also take advantage of molecular marker technologies. Studies related to genetic conservation will often rely heavily on marker studies, but will also make use of quantitative and physiological traits.

A survey of genetic diversity studies in Southeast Asia reveals two major shortcomings:

1. *Species choice*. Since most species used for such research are selected on the basis of combining economic value with ecological interest, the total number of species studied is still quite low, reflecting the limited number of species used for commercial planting. Many of the species used in commercial planting represent a small number of families. For example, *Acacia auriculiformis*, *Acacia mangium*, *P. macrocarpus* and *Dalbergia cochinchinensis* are all members of the legume family. Patterns of genetic variation observed for species from the same family might be expected to be more similar than for species from different families, so extrapolation to other species may be difficult. *Pinus kesiya*, being a conifer is likely to be representative of conifers in general, but not necessarily representative of co-distributed species.
2. *Geographic sample distribution*. The examples given in Boxes 4.1, 4.2 and 4.3 represent studies of rather limited portions of the species' natural distribution. The main reason for this has been the difficulty, at least until very recently, of acquiring samples from outside the country in which the research is being conducted. Countries of mainland southeast Asia suffered extensive unrest and poor political relations during the last few decades of the 20th century, making cooperation in research very difficult.

A project funded by the Centre for International Forestry Research (CIFOR) and the International Plant Genetic Resources Institute (IPGRI; now called Bioversity International) in 1995 addressed the first of these shortcomings with broad-based studies in which a diversity of species were selected, as opposed to only economically valuable species. For example, research conducted in two forest reserves in Peninsular Malaysia (Pasoh and Serting Tambahan), focused on eight species: *S. leprosula, Scaphium macropodum, Parkia speciosa, Daemonorops verticillaris, Garcinia malaccensis, Calophyllum ferrugineum* var. *oblongifolium, Labisia pumila and Tectaria singaporeana.* Of these, only two are commercially logged tree species (*S. leprosula* and *S. macropodum*) and one other species is valued for its seeds (*P. speciosa*). The results of this and related studies are summarised in Box 4.4.

Box 4.4 Impacts of Logging in Tropical Forests of Southeast Asia

The genetic structure of plant species with differences in economic importance and life history strategies was examined in two humid tropical forest types; a ridge forest which had been logged less than a year ago and a mixed dipterocarp lowland forest which had been logged more than 40 years ago.

About 30 mature individuals from each of 5–6 species in each forest type were randomly sampled for isozyme and RAPD analysis.

The reduction in basal area of the forest management unit in the ridge forest was about 56% after logging. Genetic diversity (expected heterozygosity for isozyme data and Shannon Diversity Index for RAPD data) was reduced by 5.0–23.4% after logging for all species examined. Loss of alleles for the different species ranged from 7.7 to 25.0% after logging. Loss of genetic diversity immediately after logging was higher in commercial timber species of low abundance (9.4% loss in heterozygosity and 25.0% loss in alleles for *S. leprosula*) than that in species of high abundance (5.0% loss in heterozygosity and 7.7% loss in alleles for *S. macropodum*) in the ridge forest management unit indicating the need to examine regeneration status for timber species of low abundance before permitting logging of these species. Among the non-timber species examined, the climbing palm, *D. verticillaris* exhibited the highest loss of alleles (13.8%). Extraction procedures for felled trees may need to be improved to minimise damage to climbing plants such as rattans.

No decline in genetic diversity measures of *P. speciosa, S. macropodum, S. leprosula, G. malaccensis, D. verticillaris* and *L. pumila* were detected in two regenerated stands (13.5% and 40.7% disturbance) when compared with an unlogged stand in a mixed dipterocarp lowland forest. When Lee et al. (2002b, c) re-examined *P. speciosa* and *S. macropodum*, respectively, using RAPD markers, substantial genetic erosion (i.e. 31.5% reduction for Shannon diversity, *H*) was detected in one regenerated stand for *S. macropodum* only corresponding to its extremely low tree density.

Genetic diversity in regenerated stands is generally maintained after a single low-intensity logging event. Assessment of genetic erosion should be carried out using more than one type of molecular marker if one of them shows no significant difference. Species' vulnerability to the threat of genetic erosion posed by selective logging is highly correlated with its abundance in a particular forest management unit. Tree density for the species can be a useful indicator in reflecting the risk of genetic erosion rather than the overall disturbance level based on reduction in basal area of all trees.

Subsequent studies on understanding the impact of logging on the population and genetic structure of different age cohorts (i.e. seedlings, saplings, poles and adults) of commercially valuable species in production forests in Peninsular Malaysia clearly indicated substantial loss of genetic diversity in adult/mature trees with the genetic diversity of seedlings/saplings being mostly unaffected but suffering sizeable logging damage (Wickneswari et al. 2004; Ho et al. 2004; Lee et al. 2007b). Ng et al. (2009) further showed that species with different breeding systems (outcrossing vs. apomictic reproduction) are affected differently by the same logging intensity, impacts to outcrossed being lower compared to apomicts.

At two sites in Thailand (Huai Kha Khaeng Wildlife Sanctuary and Sakaerat Biosphere Reserve), six species were studied: *Dipterocarpus obtusifolius, Shorea siamensis, P. macrocarpus, Sindora siamensis, Mytragina brunosis* and *Cycad siamensis*. Again, these represent a mixture of commercial (*Shorea siamensis* and *P. macrocarpus*) and non-commercial species and include a non-tree species (*Cycad siamensis*). Differences in the pattern of genetic diversity along disturbance gradients were related to the species reproductive ecology, and, to a lesser extent, the uses made of the species by local communities. For example, the two dipterocarp species differ in their reproductive ecology and economic use. *Shorea siamensis* is cut by local people for construction timber and is predominantly pollinated by a species of weak-flying bee, while *D. obtusifolius* is not harvested, and is pollinated by butterflies and moths. Ghazoul (1997) and Ghazoul et al. (1998) found that reproductive success of *Shorea siamensis* was dramatically affected beyond a critical distance of separation among individuals, representing a threshold of disturbance, which was not true for *D. obtusifolius*. The impact on reproductive success in *S. siamensis* was reflected also in reduced genetic diversity. Box 4.5 summarises these studies in Thailand.

Box 4.5 Impacts of Disturbance in Tropical Forests of Southeast Asia

Two pioneer tree species, *P. macrocarpus* and *Sindora siamensis*, and two non-pioneer tree species, *Shorea siamensis* and *D. obtusifolius*, and a non-tree species, *Cycad siamensis,* were studied along 3–6 gradients of disturbance at Huai Kha Khaeng Wildlife Sanctuary (HKKWS) and Sakaerat Environmental Research Station (SERS) in central Thailand. *Dipterocarpus obtusifolius* was only investigated in HKKWS because it was scarce in SERS. Cutting timber by local communities was the dominant form of disturbance at HKKWS while fire was inferred as the most important form of disturbance at SERS from the high incidence of cycads (Chaisurisri et al. 1997).

Genetic diversity decreased with increasing disturbance for *Shorea siamensis* and *Cycad siamensis* (Chaisurisri et al. 1997; Changtragoon 1997). The results concur well with level of exploitation of the species as *Shorea siamensis* is the most popular timber species among the four tree species investigated. No change or increase in genetic diversity with increasing disturbance was detected for *P. macrocarpus*, *Sindora siamensis* and *D. obtusifolius* (Chaisurisri et al. 1997). *Pterocarpus macrocarpus* and *Sindora siamensis* seem to benefit from some of disturbance while the results for *D. obtusifolius* are expected as it is poorly regarded as a timber species.

Studies on pollination biology and breeding system by Ghazoul (1997) and Ghazoul et al. (1998) revealed that reproductive success of *S. siamensis* was dramatically affected beyond a threshold of disturbance, which was not true for *D. obtusifolius*.

In general, tree and non-tree species respond differently to disturbance and level of exploitation.

There is an obvious contrast between the two types of studies reported here. In the first group (examples in Box 4.1, 4.2 and 4.3), the basic goal of the research was to characterise genetic diversity of a species across a significant part of its distribution – either for conservation or for genetic improvement purposes, or both. In the second group of studies (examples in Boxes 4.4 and 4.5), genetic diversity was assessed to determine the impacts of different types of anthropogenic disturbance. In the second type of study, there is a spatio-temporal problem that is common to all levels of the biodiversity hierarchy. In studying impacts of disturbance, the most direct approach is to examine a single site before, during and after the disturbance. This is often impossible, however, and even where possible, implies long-term research. An alternative approach is to study sites known to have experienced different disturbance histories. Such an approach offers the prospect of much more rapid results, but suffers from the need to make assumptions about pre-disturbance similarities among sites in the variables being studied.

The studies described above for Malaysia (summarised in Box 4.4) attempted to address this problem by adopting one approach at one site and the other at the second site. Thus, at Pasoh F.R., three sites in close proximity to each other and having different logging histories were sampled, while at Serting the same population was sampled before and immediately after a logging event. Results from the two research sites were very different – at Pasoh F.R., little impact was detected, while at Serting F.R. significant impacts were found, even for non-harvested species. The differences could be accounted for by differences in logging methods or by the different time scale – it is possible that there were impacts at Pasoh, which have been masked by time (Wickneswari et al. 2000; Wickneswari and Boyle 2000). Another problem with the research at Serting F.R. is that the materials used were vegetative, meaning that the detected impacts were limited to removal of individuals (through harvesting or death) during the logging event. The use of reproductive material yields more information in that embryos or newly germinated seedlings integrate impacts through several processes – physical loss of individuals from the population, disrupted reproductive mechanisms or increased post-fertilisation mortality. This was addressed in a later study in three production forests in Peninsular Malaysia viz. Serting F.R., Ulu Sedili F.R. and Panti F.R. where different age cohorts (i.e. seedlings, saplings, poles) and adults of three species were sampled before and after logging for estimation of genetic diversity measures (Wickneswari et al. 2004; Lee et al. 2007b) and forest structure (Ho et al. 2004). The results of these studies are also summarised in Box 4.4.

In another study in Central Kalimantan, Indonesia, reproductive material was sampled for *S. parvifolia* and *Eusideroxylon zwageri* – both species that are logged. Results for both species, in unlogged plots and plots that were logged 3 years and 7 years previously, indicated no impact on genetic diversity (Sudarmonowati et al. 1998; 2000). However, in the case of *S. parvifolia*, due to the recalcitrant nature of the seed and difficulties in harvesting unshed seed from such tall trees, germinating seedlings had to be used for the study, and their maternal parent inferred from their location. While this does not affect estimates of population level genetic diversity, it does limit the amount of information that can be collected from the material.

4.4 Methodological Issues

Since the improvement of genetic resources for production purposes is focused on a limited number of species, it is a relatively simple matter to ensure good species coverage. For example, Boyle et al. (1997) and Brown et al. (1997) identify only a very limited number of genera and species that are used in plantation forestry in Asia, and genetic diversity studies have been undertaken on most such species.

Studies of genetic variation for conservation are inherently more complex due to the numbers of species involved. It is impossible to undertake genetic diversity studies on more than a tiny proportion of species, which means that species selection has to be undertaken with great caution to ensure that the required studies are conducted on a representative sample.

For many groups of tropical tree species, their biology and physiology make genetic studies particularly difficult. For temperate and boreal species, among which conifers are a dominant group, a fortuitous division of reproductive tissue, which results in haploid endosperm derived purely form the maternal parent, makes many applications very simple. As conifers are very rare in tropical forests, similar studies are much more difficult. Furthermore, for many tropical species, seeds are recalcitrant, meaning that they cannot be stored. Also for species that occupy the canopy, their sheer size makes sampling of either reproductive or non-reproductive tissues very difficult.

In the case of genetic studies related to production forestry, there has been a strong emphasis on the use of exotics (FAO 1993; Brown et al. 1997). There are numerous reasons for this, including the fact that genetic studies are expensive, and for most developing countries donor support will be required, which may be linked to the promotion of exotic species. This has certainly contributed to the paucity of information on indigenous species which has begun to be addressed in the last 5 years or so (e.g. Lee et al. 2006).

The rapid development of molecular marker technologies offers opportunities, in terms of the both decreasing costs of some of the simpler technologies and increasing automation which reduces the opportunities for laboratory-specific differences. Also, the increasing amount of information on an ever larger number of species allows certain general trends to be determined which can help to reduce the amount of sampling and research required for many applications.

While use of molecular markers provides a better understanding of the genetic structure and mating systems of populations, the small number of species for which extensive studies have been undertaken means that the knowledge base is very narrow. Implications drawn from the small sample of species and applied to other species may not be valid. Furthermore, there exists a knowledge gap between these genetic markers and the expression of adaptive traits among individuals, families and populations. Lack of close cooperation between geneticists and physiologists to bridge this gap in knowledge could be a threat.

There is a general lack of adequate financial support from developing country governments for genetic studies of non-commercial or locally endemic species. International agencies on the other hand are more interested in funding genetic studies of species with wide geographic distribution. Hence, species with highly restricted distribution and undiscovered economic value may be lost before they are known.

4.5 Conclusions

The major threat to genetic viability of the Southeast Asian forest species is commercial logging and fragmentation to a lesser extent. Species' vulnerability to the threat of genetic viability posed by commercial logging is highly correlated with its abundance in a particular forest management unit. Tree density for the species can be a useful indicator in reflecting the risk of genetic viability rather the overall disturbance level based on reduction in basal area of all trees. Knowledge of the biological attributes of species including their local evolution can be used to develop field guides in harvesting operations to ensure genetic quality of residuals left behind in the post-harvest stands.

There is a general lack of understanding on structure of genetic variation and on the evolution of adaptability of affected populations. Though genetic diversity of a number of tropical tree species has been determined, the critical levels of the genetic parameters that render a population more susceptible to extinction are known for a few species only. DNA-based systems for the analysis of gene expression for the understanding of gene expression in different tissues and in response to biotic and abiotic stresses have not been deployed in genetic viability studies of tropical forest species. Genetic mapping of quantitative trait loci (QTL) will enable rapid assessment of ecosystem integrity for sustainable management of ecosystems.

Acknowledgement Comments and criticisms from an anonymous reviewer are gratefully acknowledged.

References

Ashton PS (1982) Dipterocarpaceae. Flora Malesiana I 9:237–552

Boyle TJB, Liengsiri C, Piewluang C (1990) Genetic studies in a tropical pine: *Pinus kesiya* II Genetic variation among four populations in northern Thailand. J Trop Forest Sci 3(4):308–317

Boyle TJB, Cossalter C, Griffin AR (1997) Genetic resources for plantation forestry. In: Nambiar EKS, Brown AG (eds) Management of soil, nutrients and water in tropical plantation forests. CSIRO, Canberra, pp 25–63

Brown AHD (1979) Enzyme polymorphisms in plant populations. Theor Popul Biol 15:1–42

Brown AG, Nambiar EKS, Cossalter C (1997) Plantations for the tropics – their Role, extent and nature. In: Nambiar EKS, Brown AG (eds) Management of soil, nutrients and water in tropical plantation forests. CSIRO, Canberra, pp 1–23

Bush RM, Smouse PE (1992) Evidence for the adaptive significance of allozymes in forest trees. New Forests 6:179–196

Cao CP, Finkeldey R, Iskandar ZS, Ulfah JS, Gailing O (2006) Genetic diversity within and among populations of *Shorea leprosula* Miq. and *Shorea parvifolia* Dyer (Dipterocarpaceae) in Indonesia detected by AFLPs. Tree Genet Genomes 2:225–239

Cao CP, Gailing O, Iskandar ZS, Ulfah JS, Finkeldey R (2009) Genetic variation in nine Shorea species (Dipterocarpaceae) in Indonesia revealed by AFLPs. Tree Genet Genomes 5:407–420
Chaisurisri KJ, Wungplong P, Liewlasaneeyanawin C, Boyle TJB (1997) Impacts of disturbance on genetic diversity of some forest species in Thailand. In: Wrap-up workshop of the CIFOR-IPGRI impact of disturbance project, Bangalore, India
Changtragoon S (1997) Impact of disturbance on genetic diversity of *Cycas siamensis* in Thailand. In: Wrap-up workshop of the CIFOR-IPGRI impact of disturbance project, Bangalore, India
Changtragoon S, Szmidt AE (2000) Genetic diversity of teak (*Tectona grandis* Linn. F.) in Thailand revealed by random amplified polymorphic DNA (RAPD). In: IUFRO Working Party 2.08.01 tropical species breeding and genetic resources: forest genetics for the next millennium. International Conferences Centre, Durban, South Africa
Coles JF, Boyle TJB (1999) *Pterocarpus macrocarpus* Kurz. A monograph. Center for International Forestry Research, Bogor, Indonesia
Das BL, Stephan BR (1984) Provenance trial of *Pinus kesiya* at Koraput, Orissa, India. In: Barnes RD, Gibson GL (eds) Provenance and genetic improvement strategies in tropical forest trees. Department of Forestry, CFI, University of Oxford, Oxford, pp 200–204
FAO (1993) Forest resource assessment 1990. Tropical Countries. FAO, Rome
Ghazoul J (1997) The pollination and breeding system of *Dipterocarpus obtusifolius* (Dipterocarpaceae) in dry deciduous forests of Thailand. J Nat History 31:901–916
Ghazoul J, Liston K, Boyle TJB (1998) Disturbance-induced density-dependent seed set in *Shorea siamensis* (Dipterocarpaceae), a tropical forest tree. J Ecol 86:462–473
Gregorius HR, Hattemer TT, Bergmann F, Müller-Starck G (1985) Unweltbelastung und Anpassungsfähigkeit von Baumpopulationen. Silvae Genetica 34:230–241
Hamrick JL, Linhart YB, Mitton JB (1979) Relationship between life history characteristics and electrophoretically detectable genetic variation in plants. Ann Rev Ecol Syst 10:175–200
Ho WS, Wickneswari R, Shukor MN, Mahani MC (2004) The effects of the timing and method of logging on forest structure in Peninsular Malaysia. Forest Ecol Manage 203:209–228
Hou D (2000) Leguminosae (subfamily Caesalpinioideae). In: Soepadmo E, Saw LG (eds) Tree flora of Sabah and Sarawak. Ampang Press, Kuala Lumpur, pp 119–180
Kanzaki M, Watanatabe M, Kuwahara J, Kendawang JJ, Lee HS, Serzawa S, Yamakura T (1996) Genetic structure of *Shorea macrophylla* in Sarawak, Malaysia. Tropics 5(1/2):153–160
Kertadikara A, Prat D (1995) Gene diversity study based on isozyme analysis of teak (*Tectona grandis* L.f.) provenances. In: Proceedings of a IUFRO symposium, Chiang Mai, Thailand, 1994, pp 227–235
Ledig FT (1988) The conservation of diversity in forest trees. BioScience 38:471–479
Lee SL (2007) Conservation strategies of a common dipterocarp endemic to Peninsular Malaysia: *Neobalanocarpus heimii* (chengal). Technical report to Bioversity International for LOA No. APO05/016, 32
Lee SL, Wickneswari R, Mahani MC, Zakri AH (2000a) Genetic diversity of a tropical tree species, *Shorea leprosula* Miq. (Dipterocarpaceae), in Malaysia: Implications for conservation of genetic resources and tree improvement. Biotropica 32(2):213–224
Lee SL, Ang KC, Norwati M (2000b) Genetic diversity of *Dryobalanops aromatica* Gaertn. F. (Dipterocarpaceae) in Peninsular Malaysia and its pertinence to genetic conservation and tree improvement. Forest Genet 7(3):209–217
Lee SL, Wickneswari R, Mahani MC, Zakri AH (2001) Comparative genetic diversity studies of *Shorea leprosula* Miq. (Dipterocarpaceae) using RAPD and allozyme markers. J Trop Forest Sci 13(1):202–215
Lee SL, Ng KKS, Saw LG, Norwati A, Siti Salwana MH, Lee CT, Norwati M (2002a) Population genetics of *Intsia palembanica* (Leguminosae) and genetic conservation of virgin jungle reserves in Peninsular Malaysia. Am J Bot 89(3):447–459
Lee CT, Wickneswari R, Mahani MC, Zakri AH (2002b) Effect of selective logging on the genetic diversity of *Scaphium macropodum*. Biol Conserv 104:107–118

Lee CT, Wickneswari R, Mahani MC, Zakri AH (2002c) Maintenance of genetic diversity in *Parkia speciosa* in logged-over forests. J Trop Forest Sci 14(2):163–178

Lee SL, Tani N, NG KKS, Tsumura Y (2004) Characterization of 15 polymorphic microsatellite loci in an endangered tropical tree *Hopea bilitonensis* (Dipterocarpaceae) in Peninsular Malaysia. Mol Ecol Notes 4:147–149

Lee SL, Ng KKS, Saw LG, Lee CT, Norwati M, Tani N, Tsumura Y, Koskela J (2006) Linking the gaps between conservation research and conservation management of rare dipterocarps: a case study of *Shorea lumutensis*. Biol Conserv 131:72–92

Lee CT, Lee SL, Ng KKS, Salwana HS, Norwati M, Saw LG (2007a) Allozyme diversity of *Koompassia malaccensis* (Leguminosae) in Peninsular Malaysia. J Trop Forest Sci 19(2):73–78

Lee KS, Wickneswari R, Choong CY (2007b) Stand structure and the genetic diversity of *Koompassia malaccensis* and *Dryobalanops aromatica* in unlogged and logged-over stands. Sains Malaysiana 36(2):233–242

Lee CT, Lee SL, Faridah QZ, Siraj SS, Ng KKS, Norwati M (2008) Genetic diversity assessment of *Koompassia malaccensis*. Pertanika J Trop Agri Sci 31(1):127–133

Liengsiri C, Yeh FC, Boyle TJB (1995) Isozyme analysis of a tropical forest tree, *Pterocarpus macrocarpus* Kurz., in Thailand. Forest Ecol Manage 74:13–22

Lim LS, Wickneswari R, Lee SL, Latiff A (2002) Genetic variation of *Dryobalanops aromatica* Gaertn. F. (Dipterocarpaceae) in Peninsular Malaysia using microsatellite DNA markers. Forest Genet 9(2):127–138

Müller-Starck G (1985) Genetic differences between "tolerant" and "sensitive" beeches (*Fagus sylvatica* L.) in an environmentally stressed adult forest stand. Silvae Genetica 34:241–247

Müller-Starck G, Baradat P, Bergmann F (1992) Genetic variation within European tree species. New Forests 6:23–47

Murawski DA, Bawa KS (1994) Genetic structure and mating system of *Stemonoporus oblongifolius* (Dipterocarpaceae) in Sri Lanka. Am J Bot 81:155–160

Namkoong G (1991) Biodiversity – issues in genetics, forestry and ethics. Forestry Chron 68: 438–443

Namkoong G, Boyle TJB, Gregorius HR, Joly HI, Savolainen O, Ratnam W, Young AJ (1996) Criteria and indicators for impacts of forest management on genetic variation. In: Testing criteria and indicators for sustainable management of forests. CIFOR Working Paper 10, Bogor, Indonesia

Namkoong G, Boyle TJB, El-Kassaby YA, Palmberg-Lerche C, Eriksson G, Gregorius HR, Joly HI, Savolainen O, Wickneswari R, Young AJ, Zeh-Nlo M, Prabhu R(2002) Criteria and indicators for sustainable forest management: assessment and monitoring of genetic variation. In: Working Paper FGR/37E, Food and Agricultural Organisation of the United Nations, Rome, Italy

Ng KKS, Lee SL, Ueno S (2009) Impact of selective logging on genetic diversity of two tropical tree species with contrasting breeding systems using direct comparison and simulation methods. Forest Ecol Manage 257:107–116

Piewluang C, Liengsiri C, Boyle T (1997) Genetic variation of *Pterocarpus macrocarpus* Kurz: variation in fruit and seed size and quality within and among natural populations. In: ASEAN Forest Tree Seed Centre Technical Publication 35. AFTSC, Muak Lek, Thailand

Saw CD (1984) Propagation of padauk (*Pterocarpus macrocarpus* Kurz). Leaflet 1/83-84. Forest Research Institute, Yezin, Myanmar

Soonhuae P, Sziklai O (1994) Genetic conservation strategies for *Dalbergia cochinchinensis* Pierre: future directions and research needs. In: Proceedings of AFTSC symposium on genetic conservation and production of tropical forest tree seed, Chiang Mai, Thailand, 1993, pp 257–266

Soonhuae P, Boyle TJB, Yeh FC (1995) The population genetics of *Dalbergia cochinchinensis* Pierre and implications for genetic conservation. In: Baradat PH, Adams WT, Müller-Stark G (eds) Population genetics and genetic conservation of forest trees. SPB Academic Publishing, Amsterdam, The Netherlands, pp 371–385

Sudarmonowati E, Hartati NS, Siregar U, Boyle TJB (1998) The effect of logging on genetic diversity of *Shorea parvifolia* in Central Kalimantan, Indonesia. In: Poster paper presented at IUFRO division 2 conference on forest genetics and tree improvement, Beijing, 1998

Sudarmonowati E, Hartati NS, Siregar U (2000) Assessing genetic diversity of two different timber species (*Shorea parvifolia* and *Eusideroxylon zwagerii*) in Central Kalimantan using DNA markers. In: Proceedings of XXI IUFRO world congress volume III, Kuala Lumpur, 2000, p 69

Szmidt AE, Wang XR, Changtragoon S (1996) Contrasting patterns of genetic diversity in two tropical pines: *Pinus kesiya* (Royle ex Gordon) and *P. merkusii* (Jungh et De Vriese). J Theor Appl Genet 92:436–441

Templeton AR (1995) Biodiversity at the molecular genetic level: experiences from disparate macroorganisms. In: Hawksworth DL (ed) Biodiversity, measurement and estimation. Chapman & Hall, London, pp 59–64

Whuangplong P, Liengsiri C, Piewluang C, Boyle TJB (1992) The effect of seed weight on growth of *Pterocarpus macrocarpus* seedlings. In: ASEAN Forest Tree Seed Centre Technical Publication 22. ASEAN Forest Tree Seed Centre, Muak Lek, Thailand

Wickneswari R, Boyle TJB (2000) Impact of logging and other forms of harvesting on genetic diversity in tropical forests. In: Young A et al (eds) Forest conservation and genetics: Principles and practice. CSIRO Plant Industry, Canberra

Wickneswari R, Zawawi I, Lee SL, Norwati M (1995) Genetic diversity of remnant and planted populations of *Hopea odorata* in Peninsular Malaysia. In: Proceedings of the international workshop of Bio-Refor, Kangar, Malaysia, 1994, pp 72–76

Wickneswari R, Lee CT, Muhammad N, Boyle TJB (2000) Impact of logging on genetic diversity in humid tropical forests. In: Matyas C (ed) Forest genetics and sustainability, vol 63. Kluwer, Dordrecht, pp 171–182

Wickneswari R, Ho WS, Lee KS, Lee CT (2004) Impact of disturbance on population and genetic structure of tropical forest trees. Forest Genet 10(3–4):193–201

Chapter 5
The Importance of Animals in the Forest

Richard T. Corlett

Abstract Most species in tropical forests are animals and most of these animals are invertebrates, particularly insects. Vertebrates are much less diverse, but dominate some key ecological roles. Almost all plants in tropical lowland forests are pollinated by animals, with bees, followed by beetles and flies, most important. Most seeds are also dispersed by animals, except in the upper canopy, with birds, fruit bats, primates, and a variety of terrestrial mammals most important. Maximum routine movement distances for both pollen and seeds are most often in the range 100–1000 m. Seeds are predated before dispersal by colobine monkeys, squirrels, birds, and insects and after dispersal by rodents, birds, ants, and other insects. After germination, the dominant herbivores are insects, and insects also dominate the consumption of dead plant material. Carnivorous animals have indirect impacts on plants through their effects on the abundances and behaviors of other animals. Few forest animals survive deforestation, while hunting is the major threat to vertebrates in intact and logged forests. Failures of seed dispersal are currently the most obvious impacts of animal losses, but more subtle impacts would be easily overlooked. Restoration of seed dispersal services is potentially one of the most effective ways of enhancing forest recovery after logging and increasing resilience to climate change.

Keywords Conservation · Decomposition · Empty forests · Gene flow · Herbivory · Pollination · Seed dispersal · Seed predation

5.1 Introduction

A tradition of early specialization by biology students on either botany or zoology has provided Southeast Asia with experts on plants or animals, but rarely both, with only the former becoming foresters. This exclusively botanical focus of forestry mattered relatively little when giant trees were being mined from extensive forests with an intact flora and fauna, but the modern forester is not only required to manage disturbed and fragmented forests with a severely depleted fauna but also often expected to protect the wildlife that remains. It is no longer possible for

R.T. Corlett (✉)
Department of Biological Sciences, National University of Singapore, Singapore 117543
e-mail: corlett@nus.edu.sg

R. Wickneswari, C. Cannon (eds.), *Managing the Future of Southeast Asia's Valuable Tropical Rainforests*, Advances in Asian Human-Environmental Research 2,
DOI 10.1007/978-94-007-2175-3_5, © Springer Science+Business Media B.V. 2011

forestry to ignore animals, even if timber production is still the main aim. This chapter explains why and then identifies both the key problems and possible solutions.

5.2 The Diversity of Animals in Tropical Forests

Most organisms in a forest are animals and there are more animal species in tropical forests than anywhere else on Earth. Insects are the most diverse group. It has been estimated that a single hectare of rainforest in Amazonia supports 100,000 species of arthropods, of which 85% are insects (Erwin 2004), while studies in New Guinea suggest that 250 herbivorous insects are associated with each rainforest tree species (Novotny et al. 2010). Other invertebrate groups with high numbers of species in tropical rainforests include the little-known mites and nematodes. Vertebrates are much less diverse than insects and other invertebrates, but they dominate in some ecological roles, such as predation on other vertebrates and seed dispersal. The best forest sites in Southeast Asia support up to 25 species of carnivorous mammals, 51 species of insectivorous bats, and 220 species of birds (Corlett 2009a).

Tree diversity is highest in the equatorial region, with a mere 52 ha of lowland rainforest at Lambir, Sarawak, supporting more tree species (1175) than the entire forests of the northern temperate zone (North America, Europe, and Asia; 1166 species) (Wright 2002). Patterns of animal diversity are not necessarily the same, although some correlation with plant diversity would be expected in the most specialized, plant-dependent groups such as herbivorous insects. Bird diversity in East Asia is highest at 25°N, where Myanmar, India, and China meet (Ding et al. 2006). This subtropical region is also very rich in plants – particularly gymnosperms – and other vertebrates, but we do not yet have enough data for most other groups of animals to know how general this pattern is.

5.3 The Ecological Roles of Animals in Tropical Forests

5.3.1 Pollination

Pollination is the transfer of pollen grains from the anther to the stigma: plant sex. Most plants require pollination to set seed, although a minority can produce seeds by self-pollination, without any external help. There appears to be a strong selection against inbreeding in tropical trees, with the result that outcrossing and long-distance dispersal of pollen are the norm. In Southeast Asia, wind pollination is largely confined to open sites and to the low-diversity forests in the north and above 1000 m altitude. Almost all tropical lowland forest trees are pollinated by animals (Corlett 2004). Bees are the most important pollinators, followed by beetles and flies, but some species depend on birds or bats. Most tree species are visited by a variety of effective pollinators, but a few species with specialized flowers or inflorescences, such as the figs (*Ficus* spp.), can only be pollinated by a single species. Pollen transfer distances vary from meters to kilometers, but most pollen in tropical forests is probably transferred tens to hundreds of meters.

5.3.2 Seed Dispersal

Pollen must be moved to the stigma of another plant of the same species, but the target for seed dispersal is much less clearly defined. Dispersal away from the parent plant reduces competition with both the parent and the siblings, as well as reducing the impact of species-specific pests and diseases. Dispersal to a site suitable for germination and growth would be advantageous, but it is usually difficult to predict in advance which site is most suitable and most plants seem to simply disperse their seeds as widely as possible. Wind dispersal of seeds is more common than wind pollination, but is largely confined to the upper canopy, where all dipterocarps and many legumes are wind dispersed, and to tiny-seeded orchids. The large, wind dispersed, seeds of dipterocarps and other canopy giants fall mostly near the parent tree, but are occasionally carried much further by short-lived bursts of high wind speed. Orchid seeds, in contrast, regularly travel much further, as shown by the presence of many orchids but no dipterocarps on the Indonesian island of Krakatau, which was sterilized by a volcanic eruption in 1883.

Most seeds are dispersed by animals, with birds, fruit bats, and primates, along with a variety of terrestrial mammals, all having a significant role in the dispersal of trees (Corlett 1998). Birds have no teeth and usually swallow fruits whole, so the largest fruit that a bird can handle is determined by its maximum gape width, which is generally related to body size. Seed dispersal distances for fruits eaten by birds depend largely on the ranging behavior of the bird species, which is also related to body size. Thus the ubiquitous bulbuls consume only relatively small fruits (<15 mm diameter) and disperse most seeds modest distances (<500 m), while imperial pigeons, hornbills, and other large frugivorous birds can move seeds from large fruits (20–40 mm diameter) for several kilometers (Table 5.1) (Corlett 2009b).

Bats have teeth, but carry large-seeded fruits to "feeding roosts," which are generally within 100 m of the fruiting tree, and drop the seeds there. Only the smallest seeds (<2–3 mm diameter) are swallowed, but these may then be carried many kilometers before being defecated. Flying foxes (*Pteropus vampyrus*) regularly fly between Peninsular Malaysia and Sumatra and may carry seeds over hundreds of kilometers (Epstein et al. 2009). Gibbons are excellent dispersal agents, swallowing fruits whole and dispersing seeds up to a kilometer, while macaques spit out larger seeds (>4 mm) relatively near (<100–200 m) the parent tree. Most carnivores, apart from cats, eat some fruit, with civets and bears the most important in seed dispersal. These animals swallow most seeds and routinely move them more than a kilometer, while muntjac and sambar deer achieve similar dispersal distances with the hard seeds that they regurgitate. Some rats and other rodents "scatter-hoard" large dry fruits and seeds, such as oaks and chestnuts (Fagaceae), burying them individually <50 m from the parent tree and effectively dispersing those that they do

Table 5.1 Routine maximum seed dispersal distances achievable by various combinations of plants and dispersal agents (Corlett 2009b)

Distance	Dispersal agent
0–10 m	Mechanical Ants
10–100 m	Wind (large winged fruits) Rodents Macaques (large seeds, not swallowed)
100 m–1 km	Small- and medium-sized forest passerines Fruit bats (large seeds) Most primates (seeds swallowed)
1–10 km	Large canopy birds Open-country passerine birds Small fruit bats (tiny seeds) Orangutans Carnivores, including civets and bears Most terrestrial herbivores
>10 km	Wind (tiny seeds), water Fruit pigeons Large fruit bats (tiny seeds) Elephants, rhinoceroses People

not retrieve. Finally, the largest terrestrial herbivores, elephants and rhinoceroses, can potentially move seeds from fallen fruits for several kilometers before depositing them in piles of fertilizing dung. So-called "megafaunal fruits" (>6 cm diameter) may be dispersed only by these animals.

5.3.3 Gene Flow

Only seeds can establish a new population of a plant species, but both seeds and pollen contribute to gene flow between established populations. The relative role of pollen and seeds in gene flow varies hugely between plant species, but in tropical tree communities as a whole, the roles are probably of similar importance, with maximum routine movement distances of both pollen and seeds most often in the range of 100–1000 m. This contrasts with most trees in the temperate and boreal zones, where pollen dispersal by wind accounts for a larger proportion of gene flow than the movement of seeds.

5.3.4 Seed Predation

Ecologists refer to the consumption of seeds by animals as "seed predation" because the whole individual plant is killed. Seed predation can occur before dispersal, if an animal eats the unripe fruits while they are still on the parent tree, or after dispersal, when an animal finds the seeds on the ground. Colobine monkeys, squirrels, some birds, and many species of beetles and moth larvae are important pre-dispersal seed predators in Southeast Asia and can greatly reduce the number of seeds available for dispersal (Corlett 2009a). Rats and other rodents are usually the most important post-dispersal seed predators, although pigs, pheasants, ants, and a variety of other insects may also be important in some cases.

Minimization of seed predation is currently the most plausible explanation for the pattern of synchronized mast fruiting at multi-year intervals that is unique to lowland dipterocarp forests. The theory is that populations of seed predators are reduced by food shortages during the long periods with low seed production, so during a mast-fruiting episode they can eat only a small proportion of the now abundant seeds and the majority of the seeds escape. Synchronization of fruiting across large areas is necessary to prevent highly mobile seeds predators, such as wild boars, from simply moving from crop to crop. Seed predators may also contribute to the Janzen–Connell effect, discussed below, if they consume more seeds when they are found at high density near the parent tree.

5.3.5 Herbivory

After the seed has germinated and the young tree has started to grow, "predation" of the whole plant becomes less likely, but animals that eat plant parts – herbivores – remain an important influence on the ecology of forests. Most animals in the forest are herbivores and most of these are insects. Seedlings are most vulnerable, with both insects and browsing mammals contributing to the very high death rates at this stage, along with fungal pathogens and falling debris. The death rate from both insect herbivores and pathogens is at least partly controlled by the density of seedlings of the same species, leading to so-called Janzen–Connell effects, with maximum seedling survival at intermediate distances from the parent tree (Hyatt et al. 2003). Indeed, it is not just high densities of the same species that are a problem, since close relatives often share the same pests. This leads to a tendency for the seedlings that survive in a patch of forest to be less closely related than would happen by chance – a tendency known as "phylogenetic over-dispersion" (Webb et al. 2006). Herbivory continues to take a toll of older plants and even adult trees can be killed by major outbreaks of insect herbivores.

5.3.6 Decomposition

The supply of nutrients needed for plant growth in tropical forests comes mainly from the decomposition of dead plant materials, rather than external inputs or the weathering of soil minerals. The decomposition of dead plants and the release of nutrients in plant-available forms are complex processes involving leaching of soluble constituents, fragmentation by soil animals, and then chemical alteration by microbes. Experiments in which soil animals are excluded by fine-mesh bags demonstrate the importance of the fragmentation stage in accelerating decomposition (Powers 2009). Termites are probably the most important group of detritivores (consumers of dead plant material) in tropical forests, but earthworms, millipedes, beetles, cockroaches, and several other insects groups are also involved.

5.3.7 Carnivory

Tigers and other carnivores do not have a direct impact on plants, but they can have a large indirect impact. This occurs partly by limiting the numbers of those animals that do interact directly with plants – frugivores, herbivores etc. – by killing and eating them, but also by altering the behavior of the survivors. A tiger in the forest means a smaller population of more wary herbivores and thus a reduced intensity of browsing. Conversely, the total elimination of vertebrate carnivores can lead to an explosion of seed predator and browser populations and thus a reduction in plant regeneration (Terborgh et al. 2001). There is also evidence that excluding insectivorous birds and bats from tree foliage increases damage by herbivorous insects (Kalka et al. 2008).

5.4 Threats to Animals in Southeast Asian Forests

The threats to animal populations in Southeast Asia are many and varied (Corlett 2009a). Deforestation removes the habitat of all forest-dependent species, although some tolerant species can persist in the mosaic of gardens, orchards, regrowth, and forest fragments that characterize areas recently cleared by peasant farmers. Industrial cash crops, in contrast, support few forest species unless large areas of forest remain nearby. Species are lost from forest fragments however well they are protected: rapidly from small fragments (<10 ha) but slowly even from very large ones (>10 km^2). Logging on its own has a selective and largely reversible impact on sensitive species, but the indirect impacts through providing access to hunters and increasing fire risk threaten many more.

Where forest still survives, hunting is the biggest threat to most vertebrates weighing more than 1–2 kg (Corlett 2007). Hunting for subsistence may be sustainable in some circumstances, but hunting for local, regional, and global markets never is. The strongest evidence for the impact of hunting, from China to Indonesia, is the existence of large areas of suitable, but unoccupied, habitat ("empty forests") for most of the vertebrate species that are widely hunted. Among invertebrates, only honeybees (*Apis* species) are subject to widespread direct exploitation, for honey and wax. The giant honeybee, *Apis dorsata*, appears to be both particularly vulnerable to overexploitation and particularly important as a pollinator of canopy trees (Corlett 2011). This species merits the level of conservation attention normally reserved for charismatic vertebrates.

5.5 Consequences for Forests of the Decline in Animal Populations

Animal species vary widely in their vulnerability to human impacts, such as hunting, logging, and habitat fragmentation. In general, larger animals are more vulnerable, because they are the preferred prey of hunters and need more and better habitat for a viable population to survive, but there are many exceptions, with pigs and deer, for example, relatively resilient to many human impacts, while some small birds and butterflies are very sensitive. Ecological processes that depend on animals are vulnerable to human impacts on those animals, but this vulnerability depends also on how many alternative animal species can fulfill the same role. At the least vulnerable extreme are processes, such as litter decomposition, which involve many, apparently interchangeable, animal species, none of which are particularly susceptible to human actions. At the other extreme are processes, such as predation on large herbivores and the dispersal of the seeds in large fruits, which depend on a small number of animal species, each of which is threatened by hunting and other impacts.

All but the best-protected forest areas in Southeast Asia have lost most or all of their largest vertebrates (Corlett 2007), so we would expect to already see impacts on those ecological processes that require them. Unfortunately, there has not yet been any serious scientific study of the "empty forests" in this region. The best evidence for ecological impacts is the many anecdotal accounts of large fruits left uneaten on the forest floor under trees that themselves grew from seeds dispersed by elephants, rhinoceroses, primates, flying foxes, or large forest birds. Note that it does not require the local extinction of a key animal species for seed dispersal to be affected: major impacts may occur in "half-empty forests," where the surviving frugivores eat and disperse only their most preferred fruits, leaving the others to rot. Loss of seed dispersal services is expected to lead to a greater clumping of offspring and a resultant higher mortality, which in some species may reach 100%. In the longer term, this will lead to a decline in animal-dispersed tree species relative to those dispersed by wind. Dispersal failures will also make it impossible for tree species to compensate for the climatic changes predicted for the coming decades by migration to more suitable habitats (Corlett 2009b).

Dispersal failures are a clearly visible consequence of disperser loss, but impacts on pollination and other animal-dependent ecological processes are much harder to detect. Most pollinators are too small to hunt, but the declines in butterfly diversity in fragmented forests (Brook et al. 2003) suggest that small size does not necessarily protect against other threats. Figs and fig wasps are a special case, because of the specificity of these relationships, their apparent vulnerability to forest fragmentation (Harrison 2001), and the keystone role of figs in tropical rainforests (Shanahan et al. 2001). Changes in pollen flow as a result of loss of pollinator species or disruption of pollinator movements may result in inbreeding and, in the longer term, a loss in plant fitness, but this will not be obvious for decades or more. Reduced gene flow through pollen may also reduce the capacity of plant populations to adapt to climate change. In the case of the giant honeybee, *A. dorsata*, there may also be a "half-empty forest" effect, with lower nest densities leading to reduced competition, reducing bee movements, and, as a consequence, outcrossing and mean pollen dispersal distances (Corlett 2011). Large changes in the populations of seed predators and browsing ungulates can result in dramatic changes in overall seedling densities (Terborgh et al. 2001), but more subtle changes in the species composition of plant regeneration are easily overlooked. The long-term monitoring of permanent plots will eventually detect any large impacts on ecological processes, but the massive amount of work involved makes this technique impractical for routine use.

5.6 Recommendations

It would be easy, but undoubtedly ineffective, to urge an end to all clearance of primary forests, a universal adoption of reduced impact logging, and a total ban on hunting. There are, however, practical actions that can be taken, at relatively little cost, and with a relatively large impact on forest heath and resilience. The foremost of these is the control of hunting of key dispersal agents: large birds, particularly fruit pigeons and hornbills, flying foxes, primates, civets, and bears – plus elephants, rhinoceroses, and tapirs, where still present. Primates, bears, and hornbills already receive near universal legal protection, but this is often not the case for the other animals mentioned, particularly pigeons, flying foxes, and civets. Subsistence and recreational hunting tends to focus on pigs, so a ban on hunting major dispersal agents will not necessarily require a total hunting ban, although this will often be easiest to enforce. Seed dispersal is necessary during post-logging forest recovery, so protection of dispersal agents during logging operations is particularly important. Manned barriers on logging roads can exclude hunters, while permanent closure of roads after logging, by deep ditches or large logs, is essential (Meijaard et al. 2005). For species that are traded, such as bears and in some areas civets, enforcement action further along the trade chain may also be needed. Where key species have already been eliminated, it would be worth considering reintroduction if there are healthy wild populations that can be used as a source (Corlett 2009a).

A focus on protecting dispersal agents will have spin-off benefits for other animals, including carnivores, but we do not know enough in most cases to target forest management practices specifically at pollination and other key animal roles. The one exception is the giant honeybee mentioned above, which needs the same attention as the key dispersal agents. Fortunately, the available evidence suggests that what is good for the timber trees is probably good for associated invertebrates. A forest with a sustainable regeneration of valuable tree species and a visible diversity of large vertebrates is likely to be in good overall health.

Deforestation, fragmentation, hunting, logging, and fire are the most obvious current threats to animal diversity in Southeast Asian rainforests, but over the next few decades these relatively well-understood problems may be increasingly overshadowed by the impacts of climate change. Current predictions suggest a warming of 3–4°C by the end of this century, with the possibility of significant changes in rainfall, although the global climate models do not agree on this. As a result, much of lowland Southeast Asia will experience climates that have not existed in the region for at least 2–3 million years. The detailed consequences for plants and animals cannot be predicted at present, but it is likely that many species will survive only by moving to areas with a more suitable climate. In relatively flat lowland areas the distance to a cooler (and/or wetter) refuge may exceed 500 km by 2100 (Wright et al. 2009), which is only practical for the most mobile species, but the steep climatic gradients in mountainous areas mean that movements of a few kilometers may allow a species to avoid extinction. To ensure that this is practical for both animals and plants, it is essential that continuous corridors of vegetation are maintained along altitudinal gradients and that the major long-distance seed dispersal agents and pollen vectors are protected from exploitation.

Finally, given the fragmented state of much of the region's vegetation, it will often be necessary to restore forest connectivity by re-planting native forest. Diverse methods are now available for doing this (Corlett 2009a), but it is still true that exotic monocultures are much easier to establish than forests of mixed native species. All practical methods of forest restoration depend on seed dispersal to bring in additional species after the initial planting and pollen movements to reduce the risk of inbreeding, so the fauna cannot be neglected. Indeed, restoring a functioning, species-rich rainforest ecosystem would be the ultimate test of our understanding of plant–animal relationships.

Acknowledgments I have incorporated ideas from many people, but I would particularly like to acknowledge Billy Hau, John Fellowes, and David Dudgeon.

References

Brook BW, Sodhi NS, Ng PKL (2003) Catastrophic extinctions follow deforestation in Singapore. Nature 424:420–423

Corlett RT (1998) Frugivory and seed dispersal in the Oriental (Indomalayan) Region. Biol Rev 73:413–448

Corlett RT (2004) Flower visitors and pollination in the Oriental (Indomalayan) Region. Biol Rev 79:497–532

Corlett RT (2007) The impact of hunting on the mammalian fauna of tropical Asian forests. Biotropica 39:292–303

Corlett RT (2009a) The ecology of tropical East Asia. Oxford University Press, Oxford

Corlett RT (2009b) Seed dispersal distances and plant migration potential in tropical East Asia. Biotropica 41:592–598

Corlett RT (2011) Honeybees in natural ecosystems. In: Hepburn HR, Radloff SE (eds) Honeybees of Asia. Springer, Berlin, pp 215–225

Ding TS, Yuan HW, Shu G et al (2006) Macro-scale bird species richness patterns of the East Asian mainland and islands: energy, area and isolation. J Biogeogr 33:683–693

Epstein JH, Olival KJ, Pulliam JRC et al (2009) *Pteropus vampyrus*, a hunted migratory species with a multinational home-range and a need for regional management. J Appl Ecol 46: 991–1002

Erwin TL (2004) The biodiversity question: how many species of terrestrial arthropods are there? In: Lowman MD, Rinker HB (eds) Forest canopies. Elsevier, San Diego, CA, pp 259–269

Harrison RD (2001) Drought and the consequences of El Niño in Borneo: a case study of figs. Popul Ecol 43:63–75
Hyatt LA, Rosenberg MS, Howard TG et al (2003) The distance dependence prediction of the Janzen-Connell hypothesis: a meta-analysis. Oikos 103:590–602
Kalka MB, Smith AR, Kalko EKV (2008) Bats limit arthropods and herbivory in a tropical forest. Science 320:71
Meijaard E, Sheil D, Nasi R et al (2005) Life after logging: reconciling wildlife conservation and production forestry in Indonesian Borneo. CIFOR, Jakarta
Novotny V, Miller SE, Baje L et al (2010) Guild-specific patterns of species richness and host specialization in plant-herbivore food webs from a tropical forest. J Anim Ecol 79:1193–1203
Powers JS (2009) Decomposition in tropical forests: a pan-tropical study of the effects of litter type, litter placement and mesofaunal exclusion across a precipitation gradient. J Ecol 97:801–811
Shanahan M, So S, Compton SG, Corlett RT (2001) Fig-eating by vertebrate frugivores: A global review. Biol Rev 76:529–572
Terborgh J, Lopez L, Nuñez P et al (2001) Ecological meltdown in predator-free forest fragments. Science 294:1923–1926
Webb CO, Gilbert GS, Donoghue MJ (2006) Phylodiversity-dependent seedling mortality, size structure, and disease in a Bornean rain forest. Ecology 87:S123–S131
Wright SJ (2002) Plant diversity in tropical forests: a review of mechanisms of species coexistence. Oecologia 130:1–14
Wright SJ, Muller-Landau HC, Schipper J (2009) The future of tropical species on a warmer planet. Biotropica 23:1418–1426

Chapter 6
Synthesis and Options for Sustainable Management

Mohamad Ismail Shaharuddin and Ratnam Wickneswari

Abstract Sustainable management of natural forests highly depend on their ability to regenerate. Successful regeneration of forests accompanied by conservation of genetic diversity of its species is important for both short-term adaptation to environmental change and long-term impact on species and communities. Forest harvesting activities invariably cause some damages to the forest ecosystem and the surrounding environment. However, it has been observed that careful and proper planning of harvesting and the implementation of reduced impact logging practices will help reduce the severity of damage to the forest. For a practical guide for genetic conservation, it is highly recommended that the requirement for the genetic conservation be incorporated in the various stages of activities undertaken during forest harvesting.

Keywords Mother tree · Silviculture treatment · Enrichment planting · Forest inventory · Residual stand · Seedling sampling · Timber tagging · Cutting limit

6.1 Introduction

The choice of a silvicultural system for a production forest depends on a host of factors which range from economic to ecological characteristics of the forest. Economic considerations are by far the most important of the factors affecting the silvicultural decision in the tropical region. In most countries in Southeast Asia where commercial logging is carried out, some form of selection felling as opposed to a uniform system is adopted with the aim of conserving stock for future use. According to Smith (1962), "a good silvicultural system is not chosen but formulated as a solution to a specific set of circumstances."

Sustainable management of natural forests depends on their ability to regenerate. Successful regeneration of forests accompanied by conservation of genetic

M.I. Shaharuddin (✉)
Institute for Environment and Development (LESTARI), Universiti Kebangsaan Malaysia (UKM), 43600 UKM Bangi, Selangor Darul Ehsan, Malaysia
e-mail: dinshah@ukm.my

R. Wickneswari, C. Cannon (eds.), *Managing the Future of Southeast Asia's Valuable Tropical Rainforests*, Advances in Asian Human-Environmental Research 2,
DOI 10.1007/978-94-007-2175-3_6,

diversity of its species and wildlife is important for both short-term adaptation to environmental change and long-term impact on species and communities. Forest harvesting activities invariably cause some damages to the forest ecosystem and the surrounding environment. Tree felling, road construction, clearing for log landing, and the movement of heavy machineries in the forest exposes soil and damages other trees and natural regeneration. For example, felling of a single big tree could damage several medium-sized trees and many saplings. Furthermore, felling of trees on steeper slopes could result in severe damage to forest stand. However, the impact is very much dependent on the degree of disturbances and the intensity of logging operation. It has been observed that careful and proper harvest planning and the implementation of reduced impact logging practices helped to reduce the severity of damage to the forest.

Figure 6.1 shows a generalized relationship between economic benefits from the forest and genetic diversity of its component species. As intensity of use increases, genetic diversity of many species will be affected, with some species unable to tolerate disturbance and becoming locally extinct. The optimum level of management in economic terms may be substantially more intensive than the optimum for conservation of biodiversity. If, for political/policy reasons, a greater effort is considered necessary to conserve biodiversity (at either the genetic or the species level), the need for compensation is inescapable, and research is required to develop an approach for establishing appropriate levels of compensation.

For a practical guide for genetic conservation, it is highly recommended that the requirement for the genetic conservation be incorporated in the various stages of activities undertaken during forest harvesting. The guidelines will cover three main categories namely pre-harvesting activities, forest harvesting activities, and post-harvesting activities.

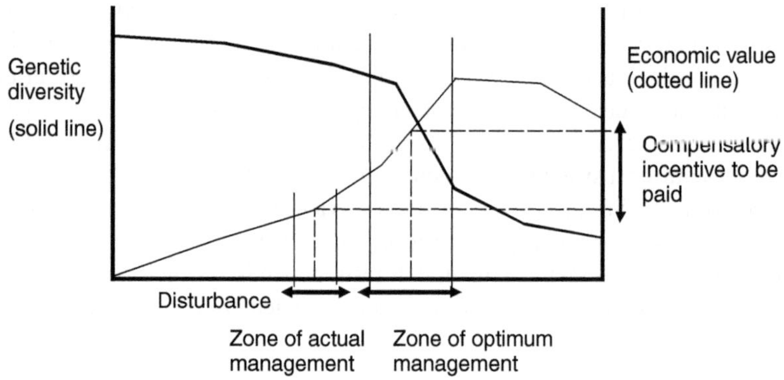

Fig. 6.1 The relationship between genetic diversity (*left axis*, *solid line*) and economic value (*right axis*, *dotted line*) as intensity of disturbance varies from low to high. The actual shape of the lines will depend on the ecological characteristics of each species and their direct and indirect uses

6.2 Pre-harvesting Activities

Forest zoning in production areas is important to classify and take into account overlapping functions. Generally, areas to be restricted include the protection of riparian buffers along rivers, which serves to protect water quality and helps to maintain wildlife corridors. High conservation value forest is designed to protect and conserve specific needs of endemic and unique species as well as for special scientific interest.

Pre-harvesting forest inventory is an inventory conducted prior to undertaking forest harvesting operations.

(1) The purpose of this inventory is to gather relevant information regarding stand stocking, species composition, and also distribution of the following tree classes for the purpose of determining optimum cutting limits. Generally, the tree size classes are enumerated as follows:

Size classes (dbh)	Class
>45 cm	Big tree
30–45 cm	Small tree
15–30 cm	Poles
5–15 cm	Small poles
1.5–5 cm	Sapling
15 cm to 1.5 m height	Seedling

(2) The inventory normally includes other information on species like bamboo, rattan, palms, and climbers.
(3) The infestation of climbers is also recorded to facilitate silvicultural decision.
(4) Preparation of stock map includes not only big trees but also distribution of natural regeneration.
(5) Good understanding of the biological attributes and ecological status of the commercially and ecological important species (described in detail in Chapters 2, 3, and 4)

Cutting limit calculation:

(1) Based on the inventory result, cutting limits for the harvesting area are determined based on size classes and species.

Tree marking and tagging:

(1) Tree marking activities involve the marking of trees for felling, mother trees for retention, and trees for protection as well as marking the direction of felling which is conducted together with tree marking activities.

(2) *Marking of trees for felling.* All trees with diameter equal to or above the prescribed cutting limits for felling are identified and marked with tagging system.
(3) *Marking of mother trees for retention.* Selected trees with good quality bole and crown are to be marked and tagged as mother trees. Principally, at least four trees below the prescribed cutting limits and with minimum diameter of 30 cm dbh, preferably well distributed over an area of 1 ha, are marked and tagged with information of serial number, tree species, dbh measurement, and tree location.
(4) *Marking of trees for protection.* Trees with diameter equal to or greater than the prescribed cutting limits are marked and tagged as protected trees. These trees are required to be retained for silvicultural purposes, sources of food and shelter for wildlife, and for the protection of environmentally sensitive areas. The information recorded are serial number of tag, tree species, dbh measurement, and tree location.
(5) *Marking the direction of felling.* Marking the direction of felling is to minimize damage to the felled tree, protected trees, mother trees, and the potential crop trees for the next harvest, as well as other residual trees. The direction of felling is marked on the trees to be felled. The criteria to be considered when marking the direction of felling are safety of fellers, minimize damage to felled trees, ease of winching and skidding, avoid damage to residual trees especially sapling mother trees and trees marked for protection, and avoid felling of trees toward protected areas and buffer strips of permanent watercourses.

6.3 Forest Harvesting Activities

Forest harvesting activities are carried out to ensure efficient forest harvesting by the logging contractor and to reduce damage to the environment and residual stand.

(1) Road construction should be guided by road specification.
(2) On felling operation, light machineries are used to minimize the impact of the surrounding area.

6.4 Post-harvesting Activities

Post-harvesting activities involve the rehabilitation of feeder roads and their closure, skid trails, log-landings, and logging camps as well as the preparation of the closing report.

(1) Post-felling inventory is an inventory conducted after the forest harvesting operation.
(2) The purpose of this inventory is to gather relevant information regarding residual stand stocking, species composition, and also distribution of tree classes for the purpose of determining optimum silvicultural treatment.

Size classes (dbh)	Class
>45 cm	Big tree
30–45 cm	Small tree
15–30 cm	Poles
5–15 cm	Small poles
1.5–5 cm	Sapling
15 cm to 1.5 m height	Seedling

(3) Based on the post-inventory result, the type of silvicultural treatment will be determined. Silvicultural treatment consists of timber stand improvement through climber cutting. Enrichment planting is carried out in poor forest and opening. It is crucial to protect the identified mother trees and also the protection trees. *Monitoring the growth and the species composition of saplings is highly recommended.*

6.5 Concluding Remarks

The judicious and careful implementation of reduced impact logging practices will further improve quality of forest management. At the same time it will reduce negative impacts of timber harvest on the environment. This includes protection of water resources as well as reducing excessive damage to the soil and other vegetation. The protection of mother trees and natural regeneration mainly of saplings will provide the necessary mechanism for the conservation of genetic resources toward sustainable forest management.

Acknowledgment The authors would like to acknowledge senior forester Mohamed Murad for providing useful insights on harvesting operations in Peninsular Malaysia and comments on an earlier version of this chapter.

Reference

Smith DM (1962) The practice of silviculture, 7th edn. Wiley, New York, NY

Glossary

Abiotic (factor) Physical or chemical factor unrelated to biological or living organisms, e.g., climate, soil, and atmospheric pollution.

Allele One, two, or more alternative forms of a gene which determine one of the inherent traits coded by the gene.

Allozyme Isozymes (proteins) whose synthesis is usually controlled by co-dominant alleles inherited by monogenic Mendelian ratios and which show up as banding patterns in electrophoresis.

Biogeography Science that attempts to document and understand spatial patterns of biodiversity. It is the study of distributions of organisms.

Biological diversity The variability among living organisms from all sources including, inter alia, terrestrial, marine, and other aquatic ecosystems and the ecological complexes of which they are a part; this includes diversity within species, between species, and of ecosystems.

Biotic (factor) Factor of biological origins related to living organisms, e.g., insect pests and fungal, viral, or bacterial diseases.

Breeding system The system by which a species reproduces. There are several natural systems in plants: (i) outbreeding (exogamy and crossbreeding) is a mating system in which mating is between individuals less closely related than average pairs chosen from the population at random. (ii) Inbreeding (endogamy and self-breeding) is the crossing of individuals that are more closely related genetically than individuals mating at random, especially when repeated for several successive generations. (iii) Clonal reproduction. A species may use one or more of these systems.

Congeners Organisms within the same genus.

Conservation (of a resource) The actions and policies that assure its continued availability and existence.

Conservation (of genetic resources) The management of human use of genetic resources so that they may yield the greatest sustainable benefit to present generations, while maintaining their potential to meet the needs and aspirations of future generations.

Conspecific Organism within the same species.

DNA marker A distinctive, readily identifiable segment of DNA.

R. Wickneswari, C. Cannon (eds.), *Managing the Future of Southeast Asia's Valuable Tropical Rainforests*, Advances in Asian Human-Environmental Research 2,
DOI 10.1007/978-94-007-2175-3, © Springer Science+Business Media B.V. 2011

Ecosystem A community of all plants and animals and their physical environment, functioning together as an interdependent unit.

Effective population size The number of individuals in an ideal population which has the same level of genetic drift and inbreeding as the population from which it is drawn.

Endangered species Any species which is in danger of extinction throughout all or a significant portion of its range.

Endemism A taxonomic category, the natural occurrence of which is confined to a certain region and the distribution of which is relatively limited.

Forest management area A forest management area refers to the clearly defined area which is under direct management of the company undertaking forest management assessment for the purpose of timber certification.

Forest zoning Classification of forest area according to defined functions based on spatial, topographical, floral and faunal, and socio-economic information. Based on the determined functions, the forest areas are classified into productive, restrictive, and protective zones.

Fragmentation The process of transforming large continuous patches into one or more smaller patches creating areas of geographical discontinuity.

Gene A chemical unit of hereditary information that can be passed from one generation to another.

Gene bank A facility established for the ex situ conservation of individuals (seeds), tissues, or reproductive cells of plants and animals.

Gene flow Exchange of genes between populations owing to the dispersal of gametes or zygotes.

Gene pool The collection of genes in an interbreeding population.

Genetic conservation All actions aimed at ensuring the continued existence, evolution, and availability of genetic resources.

Genetic diversity The variety of genes within a particular species, variety, or breed.

Genetic erosion Gradual loss of genetic diversity.

Genetic material Any material of plant, animal, microbial, or other origin containing functional units of hereditary.

Genetic resources Genetic material of actual or potential value.

Genetic variation Variation due to the contribution of segregating genes and gene interactions.

Genotype The genetic constitution of an organism as distinguished from its physical appearance.

Genus A category of biological classification ranking between the family and the species, comprising structurally or phylogenetically related species or an isolated species exhibiting unusual differentiation.

Girdling Physical cutting or disruption of the cambial sap flow within a tree. It often kills the tree.

Haploid A single complement of chromosomes, same as a gamete (pollen or egg cell).

Heterozygosity The proportion of heterozygous individuals at a locus or of heterozygous loci in an individual.

High conservation value forests High conservation value forests (HCVFs) are those that possess one or more of the following attributes:

(a) forest areas containing globally, regionally, or nationally significant concentrations of biodiversity values (e.g., endemism, endangered species, refugia) and/or large landscape-level forests, contained within or containing the management unit, where viable populations of most if not all naturally occurring species exist in natural patterns of distribution and abundance;
(b) forest areas that are in or contain rare, threatened, or endangered ecosystems;
(c) forest areas that provide basic services of nature in critical situations (e.g., watershed protection and erosion control);
(d) forest areas fundamental to meeting the basic needs of local communities (e.g., subsistence and health) and/or critical to local communities' traditional cultural identity (areas of cultural, ecological, economic, or religious significance identified in cooperation with such local communities).

Inbreeding State of a population where mating between individuals of the same family or numerous cases of self-fertilization has occurred.

Inbreeding depression Reduced fitness in a given population as a result of breeding of related individuals. It is often the result of a population bottleneck. In general, the higher the genetic variation within a breeding population, the less likely it is to suffer from inbreeding depression.

Indigenous species Species that is native, existing, or originated naturally in a particular region or environment.

Inflorescence Cluster of flowers from the same stalk.

Introgression Gene exchange from one species to another through interspecific hybridization and repeated backcrossing.

Isozyme Multiple forms of a single enzyme.

Molecular marker A gene or DNA sequence that can be used to identify an organism, species, or strain or phenotypic trait(s) associated with it.

Natural forest Forest areas where many of the principal characteristics and key elements of native ecosystems such as complexity, structure, and diversity are present, as defined by FSC-approved national and regional standards of forest management.

Outcrossing Cross-pollination between plants of different genotypes.

Permanent reserved forest Any land constituted or deemed to have been constituted a permanent reserved forest under the National Forestry Act, 1984.

Phenotype Refers to the observed state of an individual for a trait (or character) (visible physical characteristics of an organism).

Pollinator A living organism transferring pollen, e.g., insect, bird, or bat.

Polymorphic Population whose members carry at least two different alleles for a given gene.

Population A group of individuals of the same species occupying an area and genetically isolated to some degree from other similar groups.

Quantitative trait loci (QTL) Genetic loci that contribute to the variation in a quantitative trait in a population.

RAPD marker Randomly amplified polymorphic DNA is a technique which uses polymerization of DNA in a chain reaction. Its relatively low cost makes its use adapted to analyzing large numbers of polymorphic DNA markers.

Selfing Fertilization with pollen from the same flower or plant.

Silviculture The art of producing and tending a forest by manipulating its establishment, composition, and growth to best fulfill the objectives of the owner. This may, or may not, include timber production.

Species diversity The richness of animal and plant species for a given site.

Stratigraphy Studies of rock layers and layering (stratification). It is primarily used in the study of sedimentary and layered volcanic rocks.

Transgression Presence of phenotype that is more extreme than the phenotypes displayed by either of the parents.

Vicariance Separation or division of a group of organisms by a geographic barrier, such as a mountain or a body of water, resulting in differentiation of the original group into new varieties or species.

About the Editors/Authors

Chuck Cannon graduated *magna cum laude* from Harvard College in May 1989 with a degree in biological anthropology. Undergraduate research thesis on primate behavior received highest honors (*summa cum laude*) and produced two publications in international journals. Prior to enrolling in graduate school and in collaboration with a peer, had obtained funding and designed and executed a study on logged forest in Indonesia. This study had resulted in two publications, including one in *Science* magazine. Obtained Ph.D. in botany from Duke University in 2000, studied the evolution, taxonomy, and biogeography of the tropical stone oaks (*Lithocarpus*, Fagaceae). Continued this research as a postdoctoral research associate at Duke University in collaboration with Dr. Ratnam at UKM before joining the faculty as an assistant professor in the Department of Biological Sciences at Texas Tech University (TTU) in 2002. In 2007, recruited to establish a new research group focused on ecological evolution at the Xishuangbanna Tropical Botanical Garden (XTBG), Chinese Academy of Sciences at the rank of professor. In 2008, promoted to associate professor and granted tenure at TTU. Married to Chai-Shian Kua (assistant professor in the same group at XTBG) with two daughters, Surya (8) and Quana (5).

Have conducted research on a wide range of topics focused primarily on forests and forest trees of tropical Asia. This research has been based on long-term studies of reproductive behavior, advanced GIS analyses, and several different molecular DNA techniques, including custom microarrays and next-gen sequencing technology. Publications include numerous articles in leading international journals. Contributed to several book chapters. Have completed several consultancies with Wildlife Conservation Society, The Nature Conservancy, and US AID, generating technical reports and maps of conservation areas in Indonesia. Very active peer reviewer of scientific articles for numerous international journals and grant proposals for international funding agencies.

Supervised several undergraduates and graduate students at TTU, including three M.Sc. to completion. Taught courses in botany and the evolution and classification of flowering plants at TTU. Currently supervising 1 postdoctoral researcher, 4 Ph.D., 4 M.Sc., and 1 undergraduate intern at XTBG. A member of the newly established Key Lab in Tropical Forest Ecology of the Chinese Academy of Sciences based at

XTBG. Have made more than 30 oral presentations, including plenary lectures at the 50th anniversary of XTBG and the annual meeting of the Association for Tropical Biology and Conservation, on a wide range of topics in numerous countries. Leading a new intensive field course for undergraduate and graduate students in 2009, based at XTBG, in ecology and conservation research.

Richard T. Corlett obtained a first class honours degree in natural sciences from the University of Cambridge in 1974, followed by a Ph.D. in plant ecology at the Australian National University, with fieldwork in the highlands of Papua New Guinea. He has subsequently held teaching posts at the University of Chiang Mai (1980–1982), National University of Singapore (1982–1987), and the University of Hong Kong (1988–2008). He rejoined The National University of Singapore in June 2008 as professor of terrestrial ecology. His major research interests include terrestrial ecology and biodiversity conservation in Tropical East Asia, plant–animal interactions, invasive species, and the impacts of climate change. In addition to numerous scientific papers, he is the author or co-author of several books, including *Tropical Rain Forests: An Ecological and Biogeographical Comparison*, co-authored with Richard Primack and published in 2005 by Blackwell and *The Ecology of Tropical East Asia*, published in 2009 by Oxford University Press.

Mohamad Ismail Shaharuddin obtained his first degree from Universiti Putra Malaysia (UPM) and M.Sc. degree from the Australian National University, Canberra, Australia, in 1983.

His career began in the Forestry Department of Peninsular Malaysia in February 1977 as a forest planning officer. Since then he has held several positions in the department, including as deputy director of State Forestry Department, Kelantan (Development); director of Forest Planning, Forestry Department Headquarters; state director of Forestry, Negeri Sembilan & Malacca; state director of forestry, Selangor; assistant director general (Forest Resource Management), Forestry Department Headquarters, and since 1998 he has held the position of deputy director-general of forestry in charge of forest operations. He has contributed significantly to the administration, management, and conservation of the forest resources in Peninsular Malaysia. He has also steered the department in the accreditation of the MS ISO 9000 for the core process of Sustainable Timber Production from Permanent Reserved Forests and also the implementation of Malaysian Criteria and Indicators (MC&I). He has been the country focal point for Asia-Pacific Forestry Sector Outlook Study Towards 2010 where he was responsible for the preparation of Malaysia's Country Report. He was actively involved in a number of ASEAN Forestry activities such as formation of the ASEAN Institute of Forest Management (AFIM) and ASEAN-US Watershed Project. He was also a member responsible for the preparation of the Final Review of the ASEAN-New Zealand Afforestation Project located in the Philippines.

At the international level, he was a member of the Malaysian delegation to the substantive sessions of the International Tropical Timber Council (ITTC) since 1996. He has also participated actively in various committees within the ambit of ITTC and was a member of the Bali Partnership Fund Committee and also the Freezailah's Fellowship Committee. He was involved in the preparation of the consultant report for ITTC entitled 'Experiences in the Implementation of the International Tropical Timber Agreement (ITTA) 1994'. He was the leader of the Malaysian delegation at the 15th and 16th Food and Agriculture Organization's Committee on Forestry Meeting in Rome in 2001 and 2003, respectively.

He is now a principal research fellow at the Institute for Environment and Development (LESTARI), Universiti Kebangsaan Malaysia, undertaking research on areas related to environment and forestry since his retirement from the Forestry Department.

Yoshihiko Tsumura obtained his degree in forestry and agriculture from the University of Tsukuba, Japan, in 1983, followed by M.Sc. and Ph.D. in forest genetics from Institute of Agriculture and Forestry at the same university in 1985 and 1988, respectively. He started out as a research associate at the Institute of Agriculture and Forestry, University of Tsukuba, in 1988 and since then has held various positions within the Forestry and Forest Products Research Institute, Japan, and also University of Tsukuba. He is the head of Tree Genetics Lab, Department of Forest Genetics, Forestry and Forest Products Research Institute since 2006.

His major research interests include forest genetics, molecular ecology, and conservation genetics. He has about 150 refereed publications to his credentials. He was the recipient of two awards, namely Japan Forest Breeding Award in 1997 and Japanese Forest Society Award in 2009.

Ratnam Wickneswari After completing B.Sc. honours in genetics (University of Malaya) in June 1982, obtained Ph.D. in mycology and plant pathology (University of Malaya) in June 1988. Joined Forest Research Institute Malaysia as research officer in October 1986 and promoted to senior research officer and head of Genetics Unit in 1993 and 1996, respectively. Joined Universiti Kebangsaan Malaysia as associate professor in January 1998 and promoted to professor of Plant Genetics and Biotechnology in August 2005. Married to Muniandy Nalaiah (entrepreneur) with two children, Iyngaran (22 years) and Mohan Kumar (19 years).

Involved in research on population genetics of exotic and indigenous tree species (e.g. acacias, rattans, dipterocarps, ramin, grasses) and plant genomics and breeding (acacias and rice). Supervised numerous undergraduate final year students (68), M.Sc. (24), Ph.D. (8), and postdoctoral fellows (4). Currently supervising 14 postgraduate students (6 Ph.D. and 8 M.Sc.). Teaching courses on genetics and bioethics at both undergraduate and postgraduate levels.

Conducted and coordinated many research projects with collaborations nationally and internationally. Published research findings in book chapters (2), journals (67), proceedings (90), and seminars (120). Received best publication awards (7) and excellent research awards (2). Editorial board member of Forest Genetics and chief editor of Malaysian Applied Biology Journal. Chairperson of IUFRO Working Party on Population, Ecological, and Conservation Genetics (WP 2-04-01). Head of Research Cluster on Genomics and Systems Biology, UKM.

K.M. Wong, a forest botanist and plant systematist by training, has held various research appointments with the Forest Research Institute Malaysia, Sabah Forest Research Center, and Brunei Darussalam's Forestry Department, before joining the University of Malaya in Kuala Lumpur. A specialist in the taxonomy of woody bamboos and various gentianalean groups in Southeast Asia, he has also studied a number of other plant families in the region. In conservation, he successfully instigated the establishment of Brunei's Ulu Temburong National Park and provided scientific support for establishment of various protected places in Peninsular Malaysia and Sabah.

His contact with the plant life and vegetation of Malaysia and the region helped develop an interest in biological diversity at both the landscape and the regional levels. This encouraged special inquiry into special areas of species richness and their complementarity in making up the spectacularly biodiverse Sunda continental shelf territories. Malaysian Borneo, especially the so-called Northwest Borneo biodiversity hotspot and special ecoregions such as Kinabalu and the ultramafic-rock belt of Sabah, is increasingly better documented, providing opportunity for consolidating biogeographical approaches and insights that will increase conservation understanding and, hopefully, options.

Having taught ecology of tropical forests at the university's Institute of Biological Sciences, his interest has broadened into the erosion and recovery of species diversity in disturbed environments, spurred mainly by research he supervised. This might be much less attractive in that it surveys the aftermath of disturbance rather than the pristine nature of biodiversity at its best. But it is important, given that intense disturbance often takes place in areas with the highest human densities and, at the other end of the extreme, remote sites less easily monitored. He established a rainforest theme for the continuing function of the university's Rimba Ilmu Botanic Garden. Prof. Wong is on the editorial advisory committees of various botanical journals, is author of a number of books and many scientific papers, and collaborates widely in plant systematics research.

MIX
Papier aus verantwortungsvollen Quellen
Paper from responsible sources
FSC® C105338

If you have any concerns about our products,
you can contact us on
ProductSafety@springernature.com

In case Publisher is established outside the EU,
the EU authorized representative is:
Springer Nature Customer Service Center GmbH
Europaplatz 3, 69115 Heidelberg, Germany

Printed by Libri Plureos GmbH
in Hamburg, Germany